D1515010

Frontiers in Science and Technology

A SELECTED OUTLOOK

Frontiers in Science and Technology

A SELECTED OUTLOOK

A Report by the
Committee on Science, Engineering, and Public Policy
of the
National Academy of Sciences
National Academy of Engineering
Institute of Medicine

 W. H. FREEMAN AND COMPANY
New York/San Francisco

This report was prepared at the request of the National Science Foundation, under contract No. PRM-8206308.

Library of Congress Cataloging in Publication Data

Committee on Science, Engineering, and Public Policy
(U.S.)
Frontiers in science and technology.

"Prepared at the request of the National Science Foundation"—P.
Includes bibliographies and index.
1. Biology—Addresses, essays, lectures. 2. Technology—Addresses, essays, lectures. I. National Science Foundation (U.S.) II. Title.
QH311.C638 1983 500 83-1574
ISBN 0-7167-1516-3
ISBN 0-7167-1517-1 (pbk.)

Printed in the United States of America

1 2 3 4 5 6 7 8 9 0 MP 1 0 8 9 8 7 6 5 4 3

092784

Preface

This volume, entitled *Frontiers in Science and Technology: A Selected Outlook*, is the third in a series of reports on the five-year outlook for science and technology. The volume was prepared under the chairmanship of Floyd E. Bloom, of The Salk Institute, and the aegis of the Committee on Science, Engineering, and Public Policy (COSEPUP), a joint committee of the National Academy of Sciences, the National Academy of Engineering, and the Institute of Medicine. The Committee selected the topics, arranged for peer review of each chapter, and contributed to the analysis of the wider implications flowing out of these fields. The Committee also reviewed the entire report and is solely responsible for its contents.

The outlook reports were called for in the National Science and Technology Policy, Organization, and Priorities Act of 1976. A subsequent reorganization of the Executive Office of the President assigned responsibility for preparing the outlooks to the National Science Foundation. This volume was prepared by COSEPUP, on behalf of the National Academies and the Institute of Medicine, at the request of the Foundation.

As expressed in the Act, the purposes of the outlook reports are to inform the Congress of:

- Current and emerging problems of national significance that are identified through scientific research, or in which scientific and technical considerations are of major consequence.
- Opportunities for, and constraints on, the use of new and existing scientific and technological capabilities which can make important contributions to resolving these problems.

The material contained in this report is directly responsive to this charge. It reports on eight areas of inquiry that have had and promise to have important societal impacts, and which reflect the diversity, interdependence, and excitement of contemporary science and technology. The study also contains as part of the introduction a discussion of implications for the national agenda of work in these eight areas, a feature not present in earlier outlooks. By including such a discussion, COSEPUP hopes to meet more fully the congressional intent in the legislation calling for the five-year outlooks.

This volume exists because of the efforts of many individuals. James D. Ebert, Vice President of the National Academy of Sciences, was instrumental in planning it. Primary thanks for producing this report are due to the authors and reviewers, to Philip Yeager, who contributed to the discussion of implications, and, most especially, to Floyd Bloom, the Outlook's tireless and dedicated chairman. They have produced an important and readable volume which we commend to your attention.

March 1983

George M. Low, Chairman
Committee on Science, Engineering,
and Public Policy

Contents

vii

5   Surface Science and Its Applications   113
Homer D. Hagstrum, Bell Laboratories

6   Turbulence in Fluids   137
Willem V. R. Malkus, Massachusetts Institute of Technology

Committee on Science, Engineering, and Public Policy

George M. Low, Rensselaer Polytechnic Institute (Chairman)
Solomon J. Buchsbaum, Bell Laboratories
Emilio Q. Daddario, Esq., Hedrick and Lane
Elwood V. Jensen, University of Chicago
Alexander Leaf, Massachusetts General Hospital
Gardner Lindzey, Center for Advanced Study
    in the Behavioral Sciences
J. Ross Macdonald, University of North Carolina
John L. McLucas, Communications Satellite Corporation
Elizabeth C. Miller, University of Wisconsin
George E. Palade, Yale University School of Medicine
Joseph M. Pettit, Georgia Institute of Technology
Leon T. Silver, California Institute of Technology
Herbert A. Simon, Carnegie-Mellon University
I. M. Singer, University of California, Berkeley
F. Karl Willenbrock, Southern Methodist University

Ex Officio

Frank Press, President, National Academy of Sciences
Courtland D. Perkins, President, National Academy of Engineering
Frederick C. Robbins, President, Institute of Medicine

COSEPUP Staff

Barbara L. Darr
Allan R. Hoffman

Outlook Staff

Patricia C. Armstrong
Norman Metzger
Audrey Pendergast
Marjorie Sampson

Editorial Consultants

Elizabeth Dixon
Elizabeth W. Fisher
Ruth B. Haas
Alix Levy
Kenneth M. Reese
Philip B. Yeager

Introduction:
Science, Technology,
and the National Agenda

Floyd E. Bloom
Study Chairman

Director,
Arthur V. Davis Center
for Behavioral Neurobiology,
The Salk Institute

Frontiers in Science and Technology: A Selected Outlook is the third such canvass of the prospects for several fields of science and technology. All three reports were prepared in response to a congressional request that the five-year outlooks for science and technology be examined periodically. Specifically, the Congress asked that such outlooks identify national problems to which science and technology pertain and that they outline specific opportunities and constraints in their use. The request was—and continues to be—a daunting one, owing partly to the rapidity of scientific and technological progress and to the evolution of congressional priorities. The obvious need to select certain fields of science and engineering for this analysis—and even to select specific topics within each field—simply magnifies the difficulties.

Yet, we believe that the contents of this volume, reporting on current and future work in eight growth points of American science and technology, respond directly to the congressional aim to draw more fully the relation of science and technology to the very large and kaleidoscopic agenda with which the nation must deal.

Perhaps the most important theme that emerges from this volume is the remarkable fertility of American science and technology. New knowledge has provided new ways for understanding diseases. Such fundamental research areas as surface physics and turbulence theory have yielded new technologies and new controls for ancient problems. Assuming real growth in research support, the United States can ex-

The study chairman gratefully acknowledges the critical collaboration of Norman Metzger, chief editor of the Outlook Reports, in the preparation of this report.

1

pect to remain at the forefront of virtually all scientific frontiers in the 1980's and beyond.

IMPLICATIONS—FROM GENETICS TO ROBOTICS

Each chapter harbors implications affecting the national agenda. These are indicated briefly below, albeit with the suggestion that either the summaries accompanying each chapter or, more profitably, the full chapters themselves also be read.

The Genetic Program of Complex Organisms (Chapter 1)

Until the 1970's, direct study of the genes of organisms more complicated than bacteria and viruses was constrained by lack of experimental methods. With the advent of recombinant DNA and other methods, it became possible to study directly the genes of plants and animals, including those of man. It was then feasible to isolate genes responsible for particular functions, to examine their structure, and to decipher the controls on their operation.

We are in part what our genes make us, and therefore these genetic discoveries of the last 10 years will have global effects—in new insights into how an organism develops from a single fertilized egg, how it grows, and how it ages. We are also gaining new insights into some fundamental aspects of cancer; for example, rearrangements in DNA structure may be common to diverse causes of tumors, including chemical carcinogens and viruses. This work supports a shift in emphasis from the cure to the prevention of cancer. Further, we are now moving to a new level in understanding many inheritable diseases, in mitigating their effects, and in finding new methods of treating some of these once hopeless conditions.

Commercially, the new understanding of gene structure—and the research and technology that made it possible—has led already to new manufacturing techniques for creating otherwise difficult-to-make, often unavailable, and usually expensive molecules. These include insulin for treating human diabetes without the risk of immunological side effects, growth hormones for the prevention of growth abnormalities, and interferon for the possible treatment of certain cancers.

In all, the past decade in molecular biology has been marked by accomplishment, by the creation and maturation of new techniques, and by a sharper definition of questions to be probed. The next five years will provide answers and generate more questions. Also, the new knowledge will yield imaginative solutions to longstanding problems. New and better drugs and vaccines will be manufactured; new and improved plants may be developed and the yields of existing plants raised; and the diagnoses of diseases will be improved and new treatments of some of these diseases emplaced.

The Molecular and Genetic Technology of Plants (Chapter 2)

Some plants, such as carrots, will grow readily from single cells; others, such as soybeans, will not. No one knows why. This reflects an ignorance at the fundamental level of the regulation and structure of plant genes and of the factors governing the growth of plants. Gaining such knowledge has the potential to quicken the rate of breeding new plants, to raise yields, and to extend agriculture to marginal lands.

Understanding of the genetics of plants now lags that of animal genetics, so the potential remains uncertain but still enormous. Thus, whether improved understanding of plant genetics will yield new plants useful as crops remains speculation. However, the rapid pace of work in the field, marked by increasing university–industry cooperation, will surely aid in breeding more useful agricultural plants. Improved disease resistance, salt and drought tolerance, and more efficient photosynthesis are all targets of a heightened knowledge of plant genetics and the ability to manipulate it.

We may learn how to prompt seeds to accumulate more protein of a higher quality, to increase the efficiency of nitrate use, and to control plant growth and development. As our knowledge of plant genes enlarges, so will our capacity for finding faster and less expensive ways to breed crops suited to changing demands and conditions.

Cell Receptors for Hormones and Neurotransmitters (Chapter 3)

The cells of an animal or a plant must function in an integrated manner in order for the organism to survive and reproduce. In turn, integration depends upon the ability of the cells to recognize and respond to such chemical signals as hormones and neurotransmitters. Recognition and response rely on special molecules or structures in cells, called receptors. As in the case of molecular genetics, the last decade has been marked by a rapidly growing understanding of what receptors do and how they work.

In brief, receptors enable cells to react to chemical instructions from other cells. If the receptors do not work as they should or are missing, the cell malfunctions; if receptors are occupied by certain drugs or by autoantibodies to them, natural signals may be ignored. The growing knowledge of receptors, particularly since the mid-1970's, is unfolding a succession of applications and insights for a host of possible "receptor diseases" whose causes were previously unsuspected.

Further, receptor research has led to rapid and highly specific methods of following the responses to drugs of healthy and diseased cells. The mechanisms by which cells are able to regulate their responsiveness have been demonstrated. Better guides in treating human breast cancers, which may illuminate approaches to other hormonally responsive cancers, have been developed. A more detailed knowledge of how insulin works has refined our knowledge of various forms of diabetes.

In the next several years, the exact mechanisms of receptors for specific chemical messengers will become better understood. With that, our knowledge at a molecular level of how a given hormone or neurotransmitter actually exerts its effects will be enhanced.

Such knowledge inevitably will be applied to the design of drugs; that is, to finding agents that modulate, enhance, or block the actions of selected chemical messengers at their receptors. We may expect, therefore, a new generation of drugs which are much more specific in their actions and whose pharmacology will be understood at a molecular level. Similarly, approaches to diagnosis, interpreting disease symptoms, and designing treatment strategies will be altered by our heightened awareness of how cells "talk" to each other.

Psychobiology (Chapter 4)

The nerve cells of the brain communicate by using chemical and electrical signals to process information through the complex pathways by which we act and think. The search for ways to understand the physical basis of behavior and mind is continuous, difficult, and incrementally successful. A spectrum of implications is emerging from research on the brain. These include improved understanding of psychiatric and neurological diseases, the physical basis of learning, and the effects of aging. Through the use of animals in research, we now have better knowledge of the behavioral abnormalities underlying certain aspects of depression and anxiety.

There has been substantial progress in understanding some forms of deafness, suggesting what might be done about them. Advances in sensory physiology have led to electronic devices that, in a limited way, can substitute for injured eyes and ears, as well as alleviate some forms of chronic pain. Prenatal and perinatal studies are clarifying the causes of some forms of brain damage, while postnatal studies of brain development will aid in improving the education of children and will help them to reach their full capacities.

Surface Science and Its Applications (Chapter 5)

Surface science is concerned with the first few atomic layers at the surfaces of solids and examines the effects of those layers on the rest of the solid and on other materials. All solids, natural or synthetic, have surfaces, and the character of those surfaces—their structure, their electronic nature, and the atoms that they are composed of—can influence the physical and chemical behavior of the entire solid and the interaction of the solid with other atoms and molecules and with external radiation.

With the very rapid miniaturization of electronic devices, the proportion of atoms on the surfaces of integrated circuits—and hence their importance—has increased. Atoms adhering to surfaces can form "surface molecules" having, in some cases, structures not found in the gas phase. Thus, the study of chemistry at surfaces provides new insights into the interactions of molecules—events that, despite their scientific and economic importance, are still not well understood. The chemical and petroleum-refining industries rely on catalysts that speed particular reactions; an improved understanding through surface studies of how catalysts work will translate into more economic chemical processes and also into fewer side products and, hence, fewer environmental problems.

Surface studies may lead to improved corrosion-resistant materials and corrosion-preventing coatings. Surface analytical techniques are improving our knowledge of embrittlement and structural failures. The field ramifies into unlikely areas—for example, into land management, since much of soil chemistry is really colloid chemistry, which in turn is a branch of surface science dealing with small particles at a solid–liquid surface.

Turbulence in Fluids (Chapter 6)

Moving fluids tend toward turbulence, whether they are air, blood, ocean water, the aerosols of clouds, the oil flowing in pipelines, the lithosphere of drifting continents, or galaxies. Turbulence is disordered fluid flow, and understanding what initiates and maintains it is essential to better prediction and control of turbulent events.

As a practical matter, analysis of turbulence is limited to statistical approximations of what happens in turbulent flow, as distinct from what actually happens. There are now negligibly few ways of predicting turbulence quantitatively. The result is an inability to generalize—to use with confidence—the studies of turbulence associated with a particular turbine design in designing a different type of turbine. Work under way is providing new insights into turbulence that promise deeper understanding of universally applicable features of turbulent events.

Studies of turbulence may lead to improved design of axial compressors and pumps, military and civilian aircraft, chemical reactors and mixers, and more accurate prediction of ocean currents and weather patterns. Finally, new insights into the origins of disorder in natural systems may contribute not only to understanding fluid turbulence but also to model building in economics and ecology.

Lasers (Chapter 7)

Lasers are devices in which atoms, molecules, or ions are made to radiate at frequencies corresponding to separations between discrete energy levels. The product is light of essentially a single wavelength and phase—coherent light. Such light is monochromatic, or of one color; highly directional, emerging as a narrow and focused beam; and intense. The wavelength range of laser light has widened considerably in the 25 years since the laser was invented, and many novel applications have emerged. In the next several years, both the uses of lasers and their capacities will expand. Lasers are remarkable, not only for what they have done but for what they now seem on the verge of doing in a range of important applications, extending from lightwave communications to new medical surgeries and from diagnostics to separating uranium isotopes.

Laser science and technology are now driven by a reverberating dance between device physics and need: an evident application spurring the development of a suitable laser, and a new laser prompting new uses. Examples include flow cytometry, that is, the separation of cells according to their size, shape, and reflective properties, and their further sorting according to, for example, their shape and DNA content—all within minutes. Another illustration is the mode-locked laser which enables scientists to observe events occurring in trillionths of seconds, such as the chemical changes in a chlorophyll molecule triggered by light quanta, leading in turn to an improved understanding of the initial stages of photosynthesis.

The Next Generation of Robots (Chapter 8)

Robots are computer-controlled devices that reproduce human senses, manipulations, and motions well enough to do useful work. They are now being installed in manufacturing plants and, in the next several years, major expansion of their uses and capacities will occur.

Today's robots are usually sedentary, stand-alone arms, with limited precision, little sensory capability, and little if any capacity for dealing with the unexpected; thus, they can work effectively only in highly structured industrial settings.

However, these limitations are being eroded by the application of technical expertise gathered in the past 10 years. Thus, new computer languages will enable quicker, more precise communication between people and robots. New sensors will enable robots to see, to feel, to walk, and to recognize commands. Systems integrating these newly endowed traits are being developed, leading in time to a new generation of robots that can deal with disorder—that can, for example, pick a specific piece out of a bin of jumbled parts. Eventually, robots will grow in numbers and in capacities, and will become a major factor in industrial activity and change.

WIDER IMPLICATIONS

The broad themes derived from these chapters can now be discussed more extensively. They include:

- Changing perspectives on health care.
- The faster pace of technological innovation.
- The importance to scientific and engineering progress of the effective use of new communications technologies.
- The emergence of research-based technologies.
- The need to examine federal policies in regard to research and development considered as a whole.
- The quickening dissolution of the traditional boundaries between the sciences and between science and technology.

Health Care

The nation will face the escalating costs of treating chronic diseases that increasingly incapacitate and kill Americans—heart diseases, cancers, arthritis and related disorders, metabolic diseases, and a wide and complex array of psychiatric and neurological diseases.

These diseases—successors to such acute illnesses as pneumonia, tuberculosis, infantile diarrhea, syphilis, and typhoid fever as leading killers—have causes that remain unknown and develop in patterns not sharp enough to enable invariably successful intervention. Much of the care applied to them is supportive clinical management—costly "halfway technologies."

Yet, as several chapters of this report suggest, we now have new windows into many of these disorders. The unraveling of how messages are transmitted from outside to inside the cell—and what happens when that communication system is disrupted—is in turn clarifying the nature of an increasing number of diseases—hormonal illnesses, such as diabetes, and neurological diseases, such as myasthenia gravis.

Receptors, by which cells receive messages, are constructed under genetic direction, as are the molecules that direct the synthesis, storage, and secretions of the neurotransmitters and hormones. Therefore, whatever affects genetic function radiates to receptor structure and function. In turn, this coupling of molecular and receptor biology ties into work on neurological structure, physiology, and chemistry—into psychobiology, in short.

New technologies have transformed clinical practices. For example, in gynecological surgery, the carbon dioxide laser has made an enormous social impact by changing surgical procedures that once required hospitalization into procedures that can be done during office visits, with substantial savings in costs, time, and hospital resources, as well as reduced patient trauma.

There is, at the moment, a virtual explosion of biological knowledge. This may indeed be the era of biology: an era in which the innovative use of instrumentation produced from the contributions of the physical sciences—lasers, large-scale integrated circuits, and computers, for example—is merging with fundamental insights into the nature of the living cell.

The direct implication is not that chronic diseases will fade. In the next several years, however, there may emerge wholly new techniques for treating these diseases—techniques that are not simply ameliorative but penetrate to the fundamental causes of these diseases. These techniques will be highly specific, quite complex, and will require fundamental understanding for effective application.

The Pace of Technological Innovation

A 1960's study found an average 30 years' lag between basic research and technological application. Today, the lag is much shorter, as evidenced by the accelerating creation and use of new lasers, by the quick exploitation of such new scientific insights as recombinant DNA methods, by the discovery of brain peptides, and by the creation of polymetallic cluster catalysts out of surface studies.

The pace is also evident in the emergence, within the very recent past, of a new biotechnology industry principally relying on work done in the past 10 years. Similarly, the intense industrial development of very large scale integrated circuits depends on a knitting of fundamental work done almost in parallel—that work embracing materials and solid-state science, a host of new spectroscopies, and atomic and molecular physics. The character of industrial research and development in many instances has been altered by fundamental work, again much of it within the past 10 years. For example, the pharmaceutical industry has altered its approaches to designing, creating, and testing new drugs, governed by the new understanding provided by receptor biology of how signals enter and change the behavior of cells.

The implication is that we are now in an era of quickening scientific and technological change and that the two intertwine ever more closely. Opportunities are coming so fast, and competition for markets for advanced technologies is becoming so intense, that success will depend directly on the ability to create and then to exploit the new knowledge quickly.

In such an era of rapid transformation, the structures for basic research and technological development must be dynamic and must be constantly freshened by the infusion of new and highly trained scientists and engineers, by the very best instrumentation, and by unfettered communication of fundamental knowledge.

Improving Scientific and Technical Communications

The intensifying pace of scientific advance and the increasing importance of basic research to gestating new technologies combine to require that the American research system must exploit to the fullest the new communication modes now becoming available. For that reason, we discuss below some new approaches to improving the communication of scientific and technical advances—electronic networks for rapid transmission of "mail" and data; growing numbers and capacities of electronic data bases, such as those of Chemical Abstracts or the National Library of Medicine, and electronic forms of rapid publication.

The importance of these emerging communications modes to the progress of science and technology is intensified by the increasingly interdisciplinary nature of many fields of science and technology. For example, two fields described in this report—psychobiology and surface science—are generically dependent on the ability of scientists and engineers from diverse fields to interact rapidly and extensively. Progress in these fields—and, increasingly, maintenance of excellence in all fields of science and engineering—will depend on sophisticated use of the new communication modes.

It is ironic that, despite the fact that American scientists and engineers pioneer in the development and use of the latest and most sophisticated instrumentation in their experimental work, the instruments they use to exchange information have changed relatively little. With few exceptions, scientists still rely on two major and relatively traditional forms of information exchange: (1) face-to-face communication (by traveling to national or international meetings), and (2) reading scientific publications (either directly in the office or library, or by requesting the author to send a reprint). Depending upon the scientific locale and the size of the circle in which the scientist is involved, the standard sources of information may be supplemented by seminars and letter writing. However, for the most part, the information being exchanged is anywhere from months to years behind the actual state of the experimental program at the bench or the conceptual progress shared among leading members of a field.

Such continued reliance upon printed statements of progress or upon face-to-face contacts is an unnecessary impediment to the progress and communication of science and technology. Solutions for eliminating these impediments are available and could improve the communication of science substantially. These possible improvements depend upon advances that have occurred in electronic communications.

Nevertheless, a skeptic might argue that these advances in communication (perhaps viewed as just modest improvements on the telephone) will only contribute to information overloads. A major need now is to encourage a comparable advance in "smart" computer terminals through the use of programs that can digest, scan, and analyze the scientific flow of electronically stored information, filter it for each reader's self-defined profile of interests, and then condense the data. The chapter on robots has indicated the importance of the interactions between hardware and software developments for progress in automated machine capabilities. Similar extensive research and development efforts in computer language and artificial intelligence will be required to realize the full scientific and social potential of the new electronic media.

There are several implications for the nation. How can scientists and engineers be encouraged to use the new electronic modalities innovatively and effectively? What levels of support are needed? In what fields? Overall, there will be rapid changes in the communications structures in the next several years, with major impacts on the competitiveness of American science and technology. The topic needs careful examination if sound policies are to be applied.

Research-Based Technologies

The rapidity of scientific and technological advances links directly to the emergence of new research-based technologies. These are technologies that would not exist without basic research, for which the role of basic research was not to improve but to germinate. Thus, molecular beam epitaxy could not and did not exist before the perfection of new methods and insights in surface science. The 1980's will see more advanced technologies created by research. There are hints of these in this report—totally new approaches to treating major mental illnesses relying on a molecular understanding of how the brain functions; new sensory devices for the handicapped; new tactics for clinical diagnosis and surgery made possible by new lasers; new agricultural insights flowing from a deeper understanding of plant biology, and much more versatile industrial robots.

The implication is that, more than ever, basic science will be vital to technological advance and, in turn, to better productivity and enhanced economic growth. Although basic science is not inexpensive—in 1983, the federal government will provide about $6 billion for basic research out of its total federal research and development budget of $40 billion—it still is the least costly component of technological innovation. And its value in the years ahead will be multiplied as the national economy, both its manufacturing and service sectors, is suffused by advanced technologies.

Coupling Science and Technology

The nation needs no persuasion on the contribution of science and technology to its goals. The federal government is not only the leading patron of basic research in the United States, but also, to a significant degree, involves itself in the various phases of technological innovation—through demonstration programs, through developmental projects for military needs, and through special efforts of broad potential, such as the Very High Speed Integrated Circuit program of the Department of Defense.

Recognition is growing that basic research and advanced technology are critical to future economic growth and the ability to sell in global markets. In its "vision" for the 1980's, Japan's Ministry of International Trade and Industry calls for a "technology-based" nation, with more governmental support for research and development as a centerpiece. France is raising its investments in research to deal with its economic difficulties.

The nation will have to consider whether current policies regarding research and development are sufficient to meet present and future national requirements. Whether the topic is surface science, robotics, cell receptors, or the genetics of complex organisms, there is evidence of an increasing interdependence of basic research and technological development.

The nation may need to examine mechanisms for enhancing the

symbiosis of science and technology. Basic research is conducted predominantly in the universities; much applied research and most development are carried out by industry. Overall, while industry does about 70 percent of the nation's research and development, only 4 percent of that effort goes to basic research. Although only 10 percent of all research and development is done by colleges and universities, it represents 50 percent of the nation's basic research as measured in federal obligations. Should the federal government actively catalyze the university–industry cooperation that is already occurring? Should it encourage more broad-based support for research and development? If the latter course is adopted, then where should that support go? What should be the criteria? Who will select?

Boundaries

Finally, traditional disciplinary boundaries are dissolving between the fields of science and between science and technology. As examples of this trend, consider, as described in this report, how solid-state physics has merged with materials science and with chemical engineering and computer science to produce new catalysts and microelectronic fabrication methods; how optics, solid-state physics, and cellular biology have merged in the creation of flow cytometry for analyzing cell components; and how robotics and psychobiology merge in their analysis of vision.

Fundamental studies of how a smoothly flowing fluid becomes a turbulent one devolve into work in mathematics, in physiology, in the dynamics of the atmosphere, and in galactic structure. The search to uncover the structure of cell receptors needs the recombinant DNA techniques of molecular biology. And, in turn, fundamental work in receptor biology has sparked the discovery of peptide receptors governing neurological events.

The chapter on surface science also makes the point that, increasingly, science and technology are synergistically coupled. Surface science, in its modern form, is the product of several technologies that provided high-vacuum instrumentation and tools for measuring surface events and structure; one of the results of that science is the creation of a potential new technology, molecular beam epitaxy, for building semiconducting layers 100 atoms thick, varying by one or two atoms. Molecular beam epitaxy will surely create a new generation of electronic and electrooptic devices. That advance occurred in the past 15 years, and the rapidity of the progress and application is due to the close coupling of fundamental scientific work with the development and production components of technological innovation.

In fact, instrumentation increasingly knits together science and technology—nuclear magnetic resonance, scanning electron microscopes, a panoply of lasers, new sources of synchrotron radiation, and a boggling array of other tools that are indispensable to the advances recounted in this report.

The implication of these dissolving boundaries is that they alter the question of which areas of science should have priority support. The question must be rephrased before it can be answered reasonably. Because of the strong effects that advances of knowledge in one field may have on understanding phenomena in a quite different field and the major contributions to instrumentation, technique, and methodology that flow across disciplinary boundaries, it is essential that all of the major fields of science continue at strong levels of activity and excellence. In addition to general strength, but not in substitution for it, particular "targets of opportunity" may be selected for special attention from time to time. However, in the long run, the highest priority is balanced strength and an insistence on the highest quality.

These are difficult times, and the remainder of this decade may be an era of economic stresses. The temptation is to select, to have priorities, to mark some scientific fields as critical to the nation's welfare and others as less so. There is no specific historical experience to support this. If one lesson has been learned in all of the studies of the relation of scientific knowledge to technological advancement and in turn to economic strength, it is that we will be surprised. We cannot choose selective excellence in science.

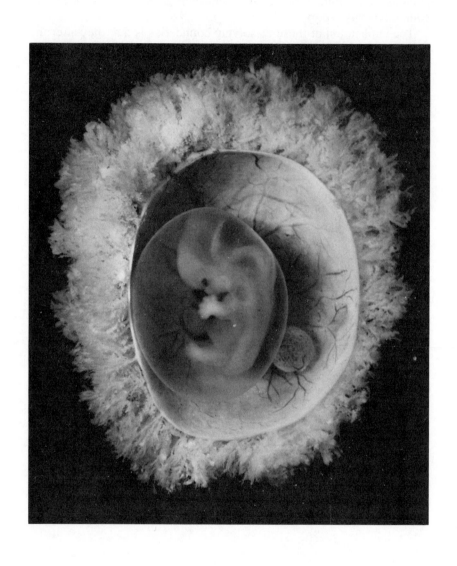

1

The Genetic Program of Complex Organisms

Myths teach us that our earliest ancestors were curious about heredity. And the history of agriculture and the domestication of animals teaches us that this curiosity served human needs. Still, although Greek thinkers stated some clear questions, it was not until the nineteenth century that the habit of careful experimentation in biology developed and the science of genetics began.

Insight into the rules of heredity is largely a product of the first half of the twentieth century. It proved enormously useful in solving basic human problems in health and agriculture even though early research was limited to what could be learned by observing the final outcome of genetic phenomena—shapes, colors, and metabolic potentials. The underlying processes remained obscure.

Then, during the 1950's and 1960's, the discovery of the structure of deoxyribosenucleic acid (DNA) and the deciphering of the genetic code showed that genetic questions are essentially chemical problems. The chemistry was, however, a difficult and complex matter and research proceeded slowly. Ten years ago, the situation changed dramatically. Now, genetic events are studied at the most basic level—that of the detailed molecular structure of genes. Already, these efforts have deepened our appreciation of the diversity and intricate interdependence of living things. They have yielded new tools for improving the

---

◀ Forty-day-old human embryo. Developmental biology seeks to understand how a single fertilized egg develops into a complex organism. [C. F. Reather, RBP, FBPA. Courtesy Carnegie Institution of Washington.]

diagnosis and treatment of diseases of plants and animals, including humans. At the same time, we have had a glimpse of how much remains to be learned if our grand hopes to understand and to use our understanding for the betterment of the planet and its inhabitants are to be achieved.

SCIENTIFIC BACKGROUND

The basic unit of all living things is a cell, be it the single cell that constitutes an entire bacterium or the millions of cells that compose a complex animal or plant. Within each cell, encoded in DNA molecules, is an information bank that contains the genetic information characteristic of the cell and of the organism. The totality of the information bank, that is, all the DNA typical of a particular organism, is called its genome, and every cell in a complex organism contains a full genome. We distinguish two major classes of organisms (figure 1): (a) eukaryotes (including all plants and animals), whose DNA is sequestered in a nucleus within the cells and (b) prokaryotes (bacteria), which have no nucleus and thus no special compartment for their DNA. A virus is something less than a cell. It can survive on its own, but it cannot multiply; to reproduce itself, it must be inside a cell. Viruses have their own genomes.

The structure of a DNA molecule may be compared to a very long but twisted stepladder with thousands to millions of rungs (figure 2). The sides of the ladder are formed of sugar molecules (deoxyribose) attached end to end through phosphate groups. To each sugar molecule is attached one of four possible molecules called bases—adenine (A), guanine (G), thymine (T), and cytosine (C). The bases on the two sides of the ladder are always paired and form the rungs of the ladder—an A on one side opposite a T on the other, and a G opposite a C. The sequence of these bases contains the genetic information. Besides providing information for the reproduction, growth, and functioning of cells, DNA contains the information needed to reproduce itself, so that precise replicas can be transmitted to new cells and offspring. These two roles of DNA are common to animals, plants, single-celled organisms, and many viruses.

How DNA Instructs Cells

Genetics is essentially the study of an information system. At the molecular level, analogies to computers are helpful. Information in DNA is encoded in a linear fashion, in the order of the four bases (A, G, T, and C): the sequence ATG means one thing, TATAAT means another, and so forth. As in a computer, the processes by which the information

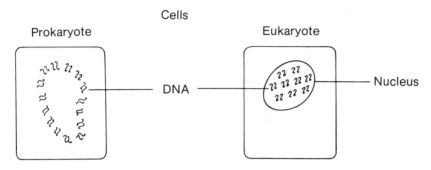

Figure 1 (Not to scale.)

is decoded are also encoded in the DNA. Computers make mistakes either by misfunctioning themselves or because of external influences—such as power shortages. Similarly, mistakes in DNA occur as a result of errors or environmental influences. But, unlike a computer, the programs for orderly decoding at appropriate times, in appropriate localities, and in suitable combinations are inherent in DNA; there is no need for instructions from outside the organism. And, unlike a computer, genetic programs include instructions for the duplication of both the information and its surrounding hardware—the cell itself. Also, DNA rearranges itself, thereby forming new combinations of information, processes, and programs. By taking advantage of both mistakes and rearrangements, DNA can try out altered programs. Some are rejected while others are adopted, often because they provide an advantage to the organism. In this way, DNA molecules are intrinsically experimental; the fruit of the experiments, over years of evolution, is the diverse natural world in which we live.

Genes Are Segments of DNA

A gene is a portion of a DNA molecule. Each gene contains a code for the structure of a single protein or single ribosenucleic acid (RNA) molecule (figure 3). RNA molecules are very like a single strand of DNA except that they contain a different type of sugar (ribose). Most genes code for proteins. The logic of the information transfer is rather straightforward. A precise copy of one strand of the DNA is made in an RNA molecule, in a process called transcription. If the gene encodes an RNA molecule, transcription is more or less the end of the decoding. But, if the gene encodes a protein, the RNA is only an intermediate, a messenger. The messenger RNA is recognized by intracellular structures called ribosomes that use the genetic code to translate the base sequence of the RNA into a specific sequence of amino acids,

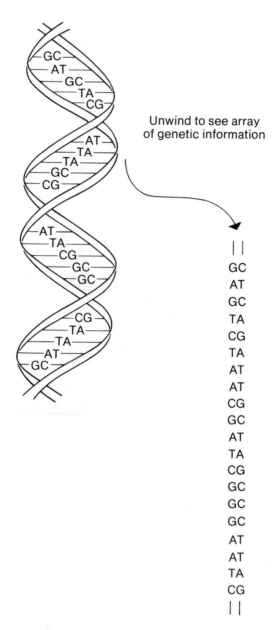

Unwind to see array
of genetic information

Figure 2

thereby constructing a particular protein. Thus, the phrase: one gene—
one protein. It is the proteins that give cells and organisms their characteristic properties, including shape, metabolic potential, color, physical capacities, and so forth. Acting as hormones, enzymes, connecting

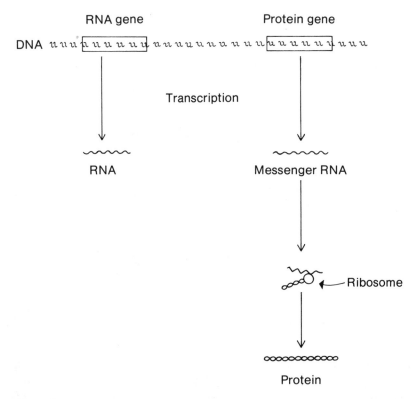

Figure 3

material, and contractile substances, proteins are the visible outcomes of the genetic program.

Not all portions of DNA molecules are genes. Some are signals that regulate the activity of nearby genes. Much of the regulation of gene activity occurs at the point of RNA synthesis (transcription). If no RNA is made, the gene remains silent. Certain DNA segments—called promoters—are signals that control the start of RNA synthesis from a neighboring gene (figure 4). A special enzyme called RNA polymerase attaches to the promoter and moves along the DNA, catalyzing RNA synthesis. At the far end of genes, special DNA sequences called terminators convey signals to stop transcription.

Molecular genetics is concerned primarily with the structure of genes and genomes, the mechanisms by which genetic information is expressed, and the ways in which that expression is regulated. A cell does not express all available genes. Different cell types (for example, liver, brain, tumor, etc.) may express different but overlapping sets of genes. Also, different genes are expressed at different times in the life

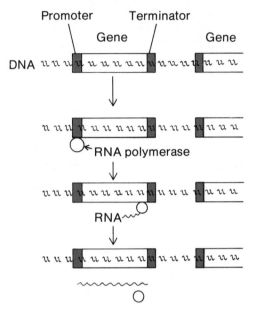

Figure 4

of a cell. During the development of a complex organism from a fertilized egg, genes are turned on and off in an orderly and programmed manner. Other genes are turned on and off in response to the cell's environment—for example, in the presence of a hormone (see chapter 3 on cell receptors). And genes are not only turned on or off; the degree to which a given gene is expressed may be modulated up or down, depending on the intracellular or extracellular environment. In addition, the gene products themselves, the protein molecules, interact with DNA in manifold ways, thereby modifying gene activity. Finally, the function of the proteins is influenced by interactions with other proteins and small molecules.

GENETIC ENGINEERING PERMITS
THE STUDY OF COMPLEX GENOMES

A great deal of what follows will be more understandable if the degree of complexity of biological systems is illustrated. The genome of a bacterium consists of one DNA molecule in a single chromosome. The bacteria *Escherichia coli*, for example, has a genome with about 4 million base pairs and about 4,000 genes. Viruses contain anywhere from 5,000 to several hundred thousand base pairs and correspondingly smaller numbers of genes. Each cell of a *Drosophila*, the vinegar fly, contains about 200 million base pairs, and mammalian cells, including those of man, contain about 3 to 4 billion base pairs. The genomes of eukaryotes are broken into several separate DNA molecules, each of which is packaged into a chromosome; human cells have 46 chromosomes. Estimates of the number of genes in a human cell vary between 10,000 to 100,000. This size and intricacy essentially prohibited investigation of the molecular genetics of complex organisms.

Until the early 1970's, molecular experimentation was, with few exceptions, limited to prokaryotes; the overall summary of gene structure given earlier in this section by and large describes prokaryote genes and their expression. Then, development of the techniques referred to as recombinant DNA, or molecular cloning, or genetic engineering, made study of eukaryotic genomes possible. With these methods, relatively short segments of DNA corresponding to single genes can be isolated free of the remainder of the genome and prepared in chemically pure form and ample quantities (that is, they can be cloned). Coincident improvements in methods for the chemical and enzymatic manipulation of DNA were equally important. First, the discovery and exploitation of a special class of enzyme made it possible to cut up the large genomic DNA into reproducible small segments. The segments are ready-made for insertion into recombinant DNA molecules. Next, it became feasible and is now common to determine with precision the sequence of thousands of base pairs on a DNA segment. Finally, there is the matter of scale; 0.000000001 (10^{-9}) grams, even 0.000000000001 (10^{-12}) grams, of DNA can be characterized with precision using the highly radioactive isotope phosphorus-32 and simple electrophoretic techniques carried out on semisolid gel supports.

Together, these methods have opened opportunities unimaginable 10 years ago. For example, it is now common to construct and store collections, or "libraries," of recombinants in which each "book" is a fragment of a genome inserted into a separate molecule. Altogether, the "library" contains the entire genome. Similarly, "libraries" containing sequences equivalent to all the messenger RNA molecules in particular cells can be obtained. Molecular cloning allows us to search within the library for a specific fragment or gene, select for it, isolate it, and amplify it to provide sufficient material for detailed analysis of its structure.

The Experimental Approach

Current research takes advantage of many different biological systems. This diversity reflects the great richness of opportunity provided by different organisms from bacteria through invertebrates, vertebrates, mammals, and primates. Viruses that infect various organisms also are used extensively both as vectors and for their intrinsic interest. The methods include the techniques of molecular biology, microbiology, genetics, virology, enzymology, and chemistry, and, most frequently, an intricate tapestry woven of all of these.

In general, the work proceeds by formulating precise questions about biological phenomena and then trying to design and carry out experiments that will answer those questions in molecular terms. Sometimes, the original plan succeeds but, just as often, unexpected phenomena are discovered and the experimental results pose new questions and even more perplexing puzzles. Frequently, the most important achievement lies in the significance of the new question, rather than any facts established. In other instances, the major achievements are the development of new methods; nothing is more vital to scientific progress than opportunities provided by imaginative technical innovations.

OVERVIEW OF PRESENT KNOWLEDGE

This past decade of research in molecular genetics has produced extraordinarily exciting results. Much of what has been learned about eukaryotes—plants and animals, including man—was totally unexpected; even the questions being asked today were unimaginable before the recombinant DNA era. The extent to which fundamental discoveries have illuminated problems in human diseases already is remarkable. Still, many of the concepts that today fuel work on human genetics arose in the past, and still arise, from studies with bacteria. And the recombinant DNA techniques themselves would not exist but for 30 years' successful work on these single-celled organisms.

Until recombinant DNA experiments revealed otherwise, many scientists assumed that eukaryotic genes were essentially similar to their familiar prokaryotic (bacterial) counterparts in structure and regulation. The genetic code is indeed universal and the general outlines of the translational processes that decode messenger RNA into proteins are also similar, as are the general principles involved in the replication of DNA. But many aspects of the molecular genetics of eukaryotes are fundamentally distinct from those of prokaryotes. No difference was more surprising than the structure of the genes themselves.

In eukaryotes, the coding region equivalent to a single gene is not a contiguous linear stretch of DNA as it is in bacteria. Rather, it is usually interrupted by noncoding stretches of DNA called introns. The entire stretch of DNA, coding regions and introns, is transcribed into a long RNA molecule; thereafter, the introns are removed from the RNA and the coding regions are spliced together to form a messenger RNA

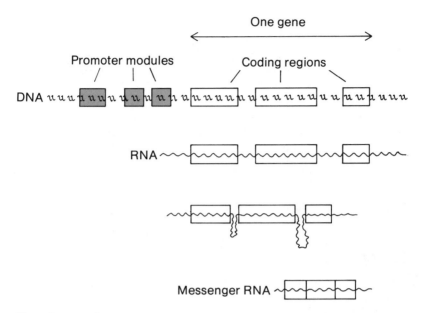

Figure 5

(figure 5). Regulatory signals that turn genes on (or off) are also distinctive in eukaryotes; these signals are complex and modular in construction and not necessarily immediately adjacent to the gene they control. Indeed, as a result of the past decade or so of work, we have learned that coding sequences of eukaryotes make up a relatively small percent of their genomes, probably 10 percent or less. There are vast stretches of DNA sequence of unknown function in between genes, within introns, and collected at centromeres (constricted regions of chromosomes). Although some genes are represented only once, others are repeated, often many times over. These redundant genes are sometimes clustered, one after another, in a long contiguous stretch of DNA. In other cases, they are dispersed to different parts of the genome, even to different chromosomes. This overall organization is in sharp contrast to the simple lining up of one gene after another in bacterial chromosomes.

Genomes Are Dynamic

Thirty years ago, genetic experiments in corn suggested that some DNA segments might be capable of movement from one place in the genome to another. Now, it is recognized that the organization of DNA is not static; rearrangements of various kinds are an inherent property of both prokaryotes and eukaryotes. Mutations, changes in DNA structure leading to altered information or altered regulatory signals, are caused not only by changes in base pairs, but also by amplifications, deletions, and inversions of DNA segments. Even insertions of DNA segments from distant genomic regions or from the DNA of infecting viruses remodel DNA structure. Mutation, then, is not necessarily the result of unprogrammed events; it can be a regular part of the system. Some rearrangements yield new combinations that are either lethal or of profound consequence (useful or detrimental) to the physiology of the cell. Other rearrangements are quite specific and have evolved for use in normal processes. These concepts emphasize that the genetic attributes of a particular species are always in flux. Nature itself provides variation as the raw material from which living systems evolve.

Genes Direct Development

The development and maturation of complex organisms from single fertilized egg cells are marvelous processes that have always been central concerns of biologists. Using the tools of molecular genetics, we are beginning to study how the developmental program operates. Different stages of development are associated with the expression of different genes. As cells differentiate to form specific organs, new sets of genes are turned on and others may be shut off. Some of the protein gene products account for the specific metabolic capabilities of given tissues. Others account for the structure of distinctive cells in, for example, the liver, blood, or brain. Still other proteins take up residence on the surface of particular cell types and probably account for the specific interactions between cells that give rise to whole organs and distinctive anatomical structures of particular shape.

Understanding Diversity

In the middle part of this century, biochemists focused on unifying notions—the structure of DNA, the universality of the genetic code, and the close similarity of metabolic pathways in diverse species. Now, the focus is shifting to the great diversity included within that unity. There is diversity in genome and gene structure, in the means for controlling gene expression, and in the way information is organized in DNA molecules. Each species is distinctive, yet even the genomes of members of a single species vary more than was ever anticipated. A sense of deep satisfaction comes from recognizing that a fundamental conviction of our society—the unique quality of each individual—is deeply rooted in the natural world.

ACTIVE AREAS OF RESEARCH*

The Structure and Expression of Eukaryotic Genes

The Structure of Genes   Figures 3 and 4 depict typical prokaryotic genes—the coding region is entirely contiguous from start to finish. Between the beginning and the end, each base pair is represented in the corresponding messenger RNA. In sharp contrast, most, but not all, eukaryotic genes are not a single contiguous stretch of DNA; rather, the coding regions are interrupted by noncoding DNA stretches called introns that are eliminated in the corresponding messenger RNA (figure 5). Coding regions, together with introns, are transcribed by RNA polymerase in the cell nucleus; by the time that the RNA arrives in the cytoplasm to be translated into a protein, the introns have been spliced out and the coding regions joined precisely to form a proper messenger RNA. In many genes, the introns are much longer than the coding regions and some genes are interrupted by 50 or more introns. It is as though sentences in a book were interrupted in the middle of words by long, irrelevant, perhaps meaningless, letters or words; the reader would need a device to recognize where the original word resumed. The cell's devices are signals designated by special base sequences that occur at the junctions between introns and coding regions. Aside from this and one or two other clues about the splicing mechanism, we remain essentially ignorant about this critical process.

One reason for intense current interest in splicing is that it represents an additional way to regulate the flow of genetic information; if the RNA is not spliced properly in a timely way, the gene is to all intents and purposes silent. Indeed, several forms of the group of inherited blood diseases called thalassemias are caused by the disruption of normal splicing. The characteristic abnormality of thalassemia is a deficiency in the production of hemoglobin, the red protein that transports oxygen in the blood. The protein portion of hemoglobin, the globin, is composed of two separate protein molecules called α and β. Thalassemias are caused by mutations in the genes for α or β. Some of these mutations are changes that lead to the production of a faulty protein, as occurs also in sickle cell disease. Others are deletions of all or part of the α or β gene. But some involve base changes at the junctions between coding regions and introns. RNA is transcribed but is not properly spliced; therefore, no messenger RNA and no protein are made.

---

*Many of the fundamental attributes of plant and animal genomes are similar. In this chapter, animals provide the examples; information on plants is found in chapter 2.

The Expression of Genes   Still other profound differences have emerged with regard to the way that gene expression is controlled in prokaryotes and eukaryotes. In the latter, special features of chromosomes affect the way that genetic information is used. In plants and animals, the long thin DNA molecules are wound by successive turnings into tight packages. At each winding, special protein molecules interact with DNA, helping to stabilize the compacted molecule. The protein-DNA complexes are termed chromatin and, in its most compact form, chromatin is visible in the microscope as the transient structures called chromosomes. But the structure of chromatin is not uniform all along DNA. In some regions, DNA is especially accessible for cleavage by enzymes. The most accessible regions are those containing genes that are being expressed. Thus, in blood cells, the genes for globin are accessible; however, in the brain, where no hemoglobin is made, the very same DNA regions resist cleavage. In pancreatic cells that synthesize the hormone insulin, regions around the insulin genes are accessible; in other tissues, where no insulin is made, these same segments of DNA resist breakdown. Expression of genes is thus associated with localized peculiarities in chromatin structure. At present, we remain ignorant about the nature of the peculiarities and the signals that account for differences among tissues. Interest is high in learning more in the coming years.

There are also major differences in the structure of prokaryotic and eukaryotic promoters. In prokaryotes (figure 4), promoter sequences sit closely flanking the coding region. Promoters for eukaryotic genes take several quite diverse forms and need not precede the gene closely. Some promoters are modular in construction and are composed of several DNA segments separated by as many as a few hundred base pairs (figure 5). Each module contributes independently to the cell's decision regarding the turning on or shutting down of a particular gene. In this way, gene expression depends on multiple signals and can be tuned very finely. Unlike anything known in prokaryotes, promoter modules for some eukaryotic RNA genes actually reside within the coding region itself (figure 6).

Ways to Study Gene Expression   In prokaryotes, a variety of regulatory proteins interact with promoter regions on DNA, thereby switching gene expression on or off. We assume that similar mechanisms exist in eukaryotes, but they have not been described on the molecular level yet. In the past few years, extracts of cellular proteins that contain active RNA polymerase have been obtained. These are beginning to yield to analysis. The next few years should see progress in understanding what proteins influence transcription and how.

Coding region

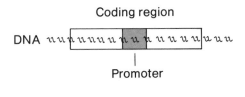

DNA

Promoter

RNA

Figure 6

Recombinant DNA methods are the basis for most studies on gene expression in complex organisms. Individual genes accompanied by their natural flanking DNA sequences are isolated in pure form and useful quantities. Also, DNA copies of messenger RNA's are prepared. Recombinant DNA techniques are then used to construct small DNA molecules containing particular genes or regulatory signals in desired arrangements. These synthetic "chromosomes" can be inserted into cells growing in laboratory dishes. The constructions are designed to answer precise questions about genetic expression. The example in figure 7 represents a circular DNA molecule of between 5,000 and 10,000 base pairs in circumference. It is made up of pieces of DNA gathered from many genomes. One segment, marked promoter, contains DNA segments needed to start messenger RNA synthesis. A bacterial gene whose gene product is a readily measurable enzyme is included, as are an intron and termination signals. A long segment marked vector is the DNA of a bacterial recombinant DNA vector and permits the whole construction to be cloned and prepared in large amounts in bacteria. Preparations of the constructed DNA molecule can be stored in the laboratory and, as required, inserted into animal cells where expression of the bacterial gene is measured. That experiment itself confirms the information used in the design—the promoter promotes, the intron is spliced out. But, more important, the construction is readily altered; each unit is removable and replaceable by recombinant DNA techniques. Other putative promoters can be tested, and other genes can be tested; the position and structure of the intron can be changed. Mutations can be introduced at will in any segment and the contribution of each base pair to function is testable. This kind of system is being used extensively to probe very specific questions regarding gene expression.

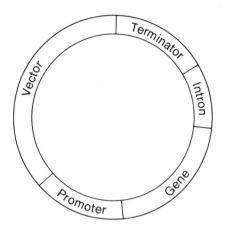

Figure 7

The Organization of DNA in Eukaryotic Chromosomes

The Organization of Genes   Unlike the single DNA molecules that make up prokaryotic genomes, eukaryotic genomes are divided into separate chromosomes. There are many more subtle differences as well. Frequently, prokaryotic genes with related functions are clustered on the DNA, and a single promoter in front of the first gene of the cluster controls expression of the entire group. This kind of arrangement is rare in complex organisms. Genes with coordinated functions are sometimes close together on chromosomes, but often they are dispersed, even to separate chromosomes. For example, the genes for the two separate globin chains, α and β, are on different chromosomes, human chromosomes 16 and 11, respectively. Yet, both must be made in the same red blood cells and at the same time if complete hemoglobin is to be produced. We do not know yet how cells coordinate the expression of dispersed genes.

Another striking characteristic of eukaryotic DNA is redundancy. Numerous DNA sequences are repeated, more or less precisely, anywhere from two to millions of times. Some of these repeated DNA segments are known to be genes, others are suspected of being genes, and still others (see below) have no known or suspected function at all.

The repetition of some genes is correlated with the need to produce large amounts of the corresponding gene product. Histones, the proteins upon which DNA is first wound in chromatin, are required by all cells and in special abundance in cells that are rapidly dividing (and thus rapidly synthesizing new DNA). Histone genes are repeated in most species; about 100 times in the *Drosophila* genome, about 10 times in humans, and over 600 times in the genome of a particular newt. The α-globin gene itself is repeated twice.

Sometimes, the copies of repeated genes are not precisely the same but differ in a few base pairs, thereby giving rise to slightly different proteins that are produced in different tissues or at different times in development. The regulatory sequences surrounding members of these gene families differ markedly, allowing for independent regulation. For example, there are in fact five globin genes of the β-type (figure 8): one is expressed early in embryonic development, two during the growth of the fetus, and two are expressed in adults. The five are clustered together on chromosome 11 and they differ somewhat from one another. Carefully timed control mechanisms, presumably associated with distinct promoter regions, assure the orderly turning off and on of these genes and thus the formation of globins suitable for various developmental stages. The nature of those mechanisms remains to be discovered.

Dispersed throughout genomes as diverse as those of *Drosophila*, yeast, mouse, and human are families of repeated sequences whose functions are unknown. Family members range anywhere from a few hundred base pairs to several thousand base pairs in length and occur between genes and in introns. At least some of them are transcribed into RNA, so they may be genes. But genes for what function? And what of those that are not transcribed? What function, if any, do they have? The amount of DNA in these sequences is astonishing; 10 percent or more of the human genome is composed of such families.

Locating Genes on Chromosomes   Classical genetic experiments assigned the locations of many genes to specific regions of particular chromosomes of *Drosophila*. But only relatively few genes could be mapped on mammalian chromosomes. Now, pure cloned DNA segments are used to locate genes on chromosomes. When chromosomes are treated with the pure segments, the segments home to the identical chromosomal DNA. This is because the two have the same base sequence and, together, can form a double-helical DNA. Using radioactive cloned segments, the location of the radioactivity is detected on the chromosomes, while individual chromosomes are distinguished by their characteristic shapes. Another useful technique utilizes separate populations of, for example, mouse cells, each of which contains a partial subset of the human chromosomes as well as all of the mouse chromosomes. Many such hybrid cells have been prepared and are grown readily in laboratory dishes. Only the cell population that includes the human chromosome carrying the gene in question will become radioactive when treated with the cloned gene segment. Often, it is possible to localize the gene in a particular region of its chromosome.

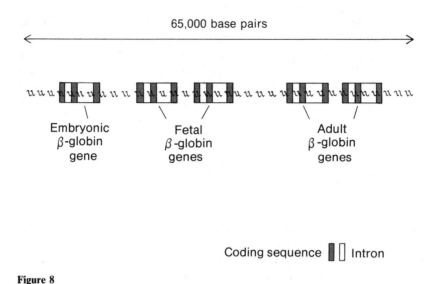

Coding sequence ▊▯ Intron

Figure 8

The list of genes whose location is known is growing daily and can be expected to increase in the next few years. This information is important for understanding the history of human chromosome construction and to aid in the diagnosis of genetic diseases associated with aberrant chromosomes.

Genomes Are Not Static

Among the programs included in DNA are some that provide for the formation of new combinations of genetic information. In recent years, we have learned that there are many different programs for rearranging DNA and that they operate much more frequently than previously supposed.

Reassortment of genetic information occurs in several ways. All cells in eukaryotes except eggs and sperm contain two copies of each chromosome; the eggs and sperm contain only one copy of each. At fertilization, a duplicate set is reinstated. The new offspring receives one chromosome of each pair from each of its parents. When the new individual matures each egg or sperm receives, at random, one member of each chromosomal pair. In this way, genetic information is constantly shuffled within a species, but new DNA molecules are not formed. Special constricted regions of chromosomes called centromeres have long been known to play a role in this sorting out process. Recently, centromeric DNA sequences were isolated from yeast. When included in appropriate recombinant DNA constructions, these

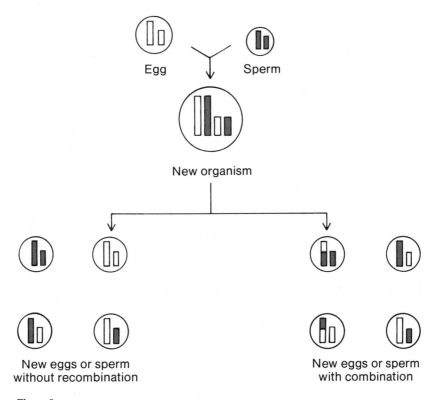

Figure 9

sequences act in yeast cells like true minichromosomes. They replicate, and the duplicates are properly sorted into two daughter cells on cell division.

The unpaired chromosomes in eggs and sperm are frequently not identical to the ones received from the original parent. During the process of maturation of germ line cells, paired chromosomes often interchange pieces (figure 9). Of course, the two members of a pair of chromosomes are overwhelmingly identical (or homologous). But, if any mutations or infrequent gene forms are present on the interchanging pieces, a new combination of genetic information is produced. This kind of interchange is called homologous recombination and its chemical mechanism depends on the near identity of the DNA molecules on pairs of chromosomes. This type of recombination was discovered early in this century and it became the primary tool of classical genetic analysis.

Figure 10 shows some additional recombination mechanisms. Deletions of DNA, amplification of particular segments to multiple copies,

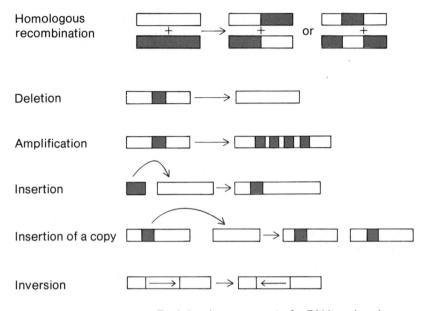

Each bar is a segment of a DNA molecule

Figure 10

insertions of a segment or a copy of a segment from a distant region of DNA, and inversion of DNA segments have been analyzed at the molecular level. However, neither in these cases nor for homologous recombination is the actual chemical mechanism known. Covalent chemical bonds must be broken and new ones formed. How does this occur? What enzymes and proteins are involved? The next few years should bring intensive work on these questions.

Movable DNA Segments   The discovery of DNA segments that move about from one place to another in genomes (by insertion) warrants special attention. The existence of movable elements was first suspected more than 30 years ago, on the basis of classical genetic experiments with corn. This property of DNA has now been demonstrated at the molecular level in bacteria, yeast, and *Drosophila*. There is good presumptive evidence that it occurs in all species, including man. Most often, special DNA sequences do the moving. Multiple copies of these DNA segments are spread throughout genomes and their own base sequence somehow bestows the ability to jump from one place to another. In some cases, it is a copy of the unit that moves, leaving the original in place. The arrival of a movable element in a new position results in a mutation if a coding sequence or a regulatory element is interrupted. In bacteria, movable elements sometimes bring along excess baggage in the form of neighboring genes. This is one way to generate duplications of a gene. Probably the same is true in eukaryotes, although the facts are not clear. This rather astonishing property of DNA is attracting much attention. Possible mechanisms to explain movement are being tested. But, most of all, we are curious to know whether the elements serve any purpose besides being rambunctious, and also how frequently they jump about.

Recombination May Play a Role in Evolution   Biological evolution is usually thought of in terms of fossils and changes in the shapes of structures, but it must reflect the evolution of genomes. Changes in DNA provide the variation upon which natural selection operates in evolution. Mutations are caused at random not only by external environmental influences such as chemical carcinogens, or X-rays, or by such internal processes as mistakes during DNA replication, but also by deletions, amplifications, and insertion of movable segments. In the past, mutations were generally detected as odd external characteristics or as faulty gene products—proteins that work poorly, or not at all, or are not made. Some mutations are of minimal consequence to the organism, but others cause lethal genetic diseases, like some of the thalassemias described earlier. Now, recombinant DNA techniques allow a direct look at the structure of DNA within and around specified genes. Genetic diseases can be diagnosed in this way. And, to our surprise, even the DNA of apparently normal individuals of a single species varies a great deal, especially in noncoding regions. In many cases, the variation is the result of deletions and amplifications of small regions of DNA. It is not easy to know what "normal" DNA is. Are the variations of some subtle consequence to the organism and to evolution? Probably, the answer will be "yes and no, depending," but the next few years surely will see increasing attention focused on the extent and significance of variation, especially in human DNA.

Examination of the structure of related genes has already shown that new genes arise by amplification of existing genes; that is, by formation of two or more copies of what was a single gene. The multiple β-globin genes (figure 8) are a case in point. Their structures are too similar to have arisen separately; the more reasonable conclusion is that they share a single common ancestor. Not all genes with a common ancestor are clustered together; some have moved to distant positions on DNA, perhaps by means of movable elements.

If there are two or more copies of an essential gene, the extra copies are free to collect mutations since the required function is maintained by one copy only. Sometimes these changes lead to a slightly altered function and, in this way, new genes can arise. There are examples beside the globins. Actin is an abundant protein required for all movement in living things, including muscle contraction and perhaps the separation of chromosomes during cell division. Four or more genes for actin occur in vertebrates. Some encode muscle actin, while the others encode slightly different actins used specifically for intracellular movement. More distant relationships can be discerned between genes for insulin and several other seemingly unrelated hormones. The fact of gene duplication is confirmed by the existence of different numbers of the same gene in different organisms; there are, for example, two insulin genes in rats and only one in humans.

Considerations like these, and others, mean that ideas about evolution are themselves evolving. Genomes can expand, contract, and rearrange information as well as be changed by simple mutations. It is easier now to imagine how genomes may have been constructed and altered, leading to the formation of new species.

DNA Rearrangements and Disease These comings and goings of DNA segments are of much more than academic interest. Increasingly, it appears that rearrangements in DNA may prove to be a common element in the diverse causes of tumors, including chemical carcinogens and viruses. Although human tumor viruses have proved elusive (several are tentatively identified), such agents are known to affect many vertebrates. After infecting a susceptible cell, the genomes of tumor viruses are inserted, like movable elements, into the cell genome (figure 11); indeed, the structural organization of the viral DNA is remarkably similar to that of some well-known movable elements.

Among the genes carried into the cell by the virus, there is frequently an oncogene. Oncogenes encode proteins that alter cellular function drastically and turn the cell into an oncogenic cell, thus starting a tumor. But other tumor viruses carry no oncogene at all and, while they are less efficient at causing tumors, they, too, insert their DNA into the cellular genome and are weakly oncogenic (figure 11). Chemical carcinogens, too, cause tumors without introducing new genes.

What is common to these different situations? Important clues may be provided by one recent discovery: the oncogenes of tumor viruses exist in cells even without the virus. They are normal cellular genes. In normal cells, it may be that expression of the oncogene is modulated to avoid excessive production of its protein product. But the invading copy on the tumor virus DNA is regulated by viral control signals, not cellular ones. An excess of the gene product would then be made and interfere with the delicate balance on which normal cell function depends. What of the tumor viruses that have no oncogene? They do contain regulatory signals. It appears that on some occasions these

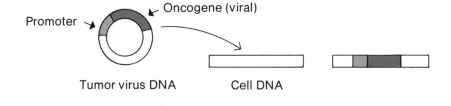

Promoter

Oncogene (viral)

Tumor virus DNA   Cell DNA

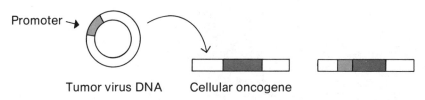

Promoter

Tumor virus DNA   Cellular oncogene

Figure 11

DNA regions are inserted in such a way that they take over the regulation of the cellular "oncogene." Similarly, it is not difficult to imagine a chemical carcinogen or X-rays contributing to aberrant regulatory signals. The net result may often be similar: excess oncogenic protein and physiological imbalance leading to a tumor cell. Still, chemical carcinogenesis and cancer itself are complex, multistep phenomena and are surely more complicated than this simple model implies.

These results provide a substantial new approach to a critical medical problem. At this stage, they raise many new questions, including how this new knowledge can be used to develop clinically useful techniques. It is also interesting to think about how a virus and a cell come to share genes. Perhaps an ancestral form of the virus recombined with cellular genes during some early infection, picking up the oncogene by chance in the process. As always, such interesting problems attract research activity, and the near future surely will bring much more information.

Using Recombination as an Experimental Tool   Animal DNA segments or replicas that have been purified and amplified by recombinant DNA techniques are taken up by animal cells growing in laboratory dishes. Inside the cells, the DNA fragments recombine with cellular DNA and become a permanent part of the genome; they are passed on to daughter cells and their genes are expressed. Appropriate genes added in this way overcome the defects in mutant genes within the

The successful transfer of a rat gene coding for growth hormone produced the "super-mouse" on the right. Moreover, the fused gene is heritable, so that big mice appear in the second generation. [Ralph D. Brinster, University of Pennsylvania.]

original cell genome. By altering promoter signals, or splicing sites in the cloned DNA, regulatory mechanisms can be investigated while the gene is resident in chromosomes. This type of experiment provides a model for approaches to therapy for genetic diseases. Already, foreign genes have been incorporated into the genomes of fertilized mouse eggs and the expression of the gene observed in the progeny (see also the discussion of the Ti-plasmid in chapter 2).

Very recently, these techniques provided important insights into tumor cells. Tumor cell DNA isolated from human or mouse tumors converts laboratory cells growing in dishes to tumor cells. The DNA segments responsible for this oncogenic transformation were purified by molecular cloning. The results so far suggest that specific DNA segments are oncogenic in different tissues. For example, the same oncogenic DNA fragment is isolated from tumors of the colon derived from different humans. Furthermore, several such DNA segments are known to be related to the oncogenes already identified in both tumor viruses and normal cells (see above). Are there many different normal genes that can turn oncogenic, or is there only a small group of such genes? What proteins are encoded by these genes? These experiments have created much excitement and have raised many new questions to be probed by the tools of molecular genetics.

Recombination in the Control of Gene Expression   The recombinational events described so far have a substantial element of randomness. The processes encoded in DNA make them possible, but whether or not a particular new arrangement of DNA molecules occurs seems, to a large extent, to be a matter of chance. Other recombinational events are specifically directed by programs in DNA and play a critical role during the life of an organism. The most dramatic, but not the only, example is the series of events that leads to the formation of antibodies, the proteins that protect (or immunize) humans and other vertebrates from foreign agents such as viruses, bacteria, and even cancer cells. For 50 years, the immune system has been a nagging puzzle. How could a genome of finite size and apparently fixed structure encode a seemingly infinite number of proteins, each a specific antibody, to some foreign material that the organism had never encountered before? Data on the structure of antibody proteins themselves gave some clues, but only when genes for antibodies were cloned and characterized did the answer emerge.

While a great many questions remain and are being studied actively, the major outlines of the explanation can be stated. In essence, the genes for formation of antibodies do not exist as such in early embryos. Rather, the several DNA segments needed to construct antibody genes are present but are not contiguous. Also, there are multiple but not identical copies of some of the segments. Later in development, the dispersed segments are joined together by recombination in those cells (lymphocytes) that manufacture antibodies. Each developing lymphocyte assembles a different set of DNA segments, thereby constructing a distinctive gene. It then multiplies to form a clone of cells, each of which can produce the same distinctive antibody. Many different families of lymphocytes patrol the organism, each providing protection against a different foreign invader. Three or four different coding regions are needed to assemble the gene for each of the two parts of an antibody protein. Given the multiple but not identical copies of these regions, many different combinations are possible. It is estimated that about 18 billion different antibodies (in mice, and a similar number in humans) can be made. This extraordinary discovery tells us two important things. First, the abundance of specific antibody responses is no longer a mystery; with a relatively small amount of DNA devoted to the task, vertebrates are prepared to combat almost any invader. Second, the old dogma that the DNA in every cell of a complex organism is exactly like the DNA in the fertilized egg from which it arose is incorrect, at least for lymphocytes. Similar programs for reshuffling genes are used in other organisms.

An interesting example is the way in which trypanosomes evade the immunological defenses of the humans and domestic animals they infect. Trypanosomes are protozoan parasites widely prevalent in Africa that are spread by tsetse flies. Like some other serious parasitic diseases, trypanosomiasis has a cyclical character; the disease repeatedly flares up, quiets down, and then flares up again. At first, antibodies to the protein that covers the surface of the trypanosome are formed by the immune system and infection is conquered. Then, a new group of trypanosomes multiplies. The latter trypanosomes are unaffected by the prior antibodies because they have disguised themselves with a new protein coat. How? There is a family of more than 100 genes that encode different coat proteins and, when one no longer protects the parasite, it switches over and uses a different gene for a new coat protein. The switch involves moving the new gene (or a copy thereof) into a distinct and active position in the genome. These genes, like cassette tapes, are normally quiet when held in storage; when they are moved to the active position (like a tape player), they are turned on. At a later time, the active gene is replaced by a new insertion and the process is repeated. A related mechanism determines the mating-type (or "sex") of yeast cells.

These examples show that evolution has put recombinational programs to work in normal processes. Natural selection is opportunistic, and we should not be surprised at the profusion of mechanisms it adopts. How frequently is recombination used to regulate gene expression? There are thousands of programs and organisms that are uninvestigated, and each has evolved to meet the challenges of its environment. We can expect that they exploit fully the opportunities provided by recombination.

Development and Differentiation

Developmental biology seeks to understand the mechanism by which a single fertilized egg develops in an orderly and programmed manner into a complex eukaryote constructed of individual, highly specialized cells and tissues. Bees build beehives and men build airplanes, but these constructions are almost trivial compared to the complex job of building a living organism. Timing is a critical factor in development;

each event must follow properly those that precede it if the final product is to be correct. Place is also essential; specific metabolic capacities are called forth in particular tissues, and anatomical features develop in precise locations. One of the outcomes of development is differentiation—the specialization of some cells to perform certain jobs in certain places. Brain cells are needed in the brain, liver cells in the liver, light sensitive cells in the eye, and so forth. As with so many other problems in biology, molecular genetics is being used to probe decades-old, even centuries-old questions about development and differentiation.

DNA encodes the developmental program. Known mutations are associated with the cessation of development or with peculiar differentiation. *Drosophila*, which reproduces rapidly and develops quickly from external eggs in well-defined stages, is an ideal organism for these studies. Sea urchins and, among vertebrates, the toad *Xenopus* are also convenient experimental animals.

The overall established principle is that some genes are turned on or off or modulated at precise times during development or in some particular differentiated tissue. We have already described the five β-globin genes and how they are expressed at different stages in human development. The story of antibody gene assembly illustrates a process of differentiation; the lymphocytes become highly specialized factories for producing a single type of antibody. In cells that are differentiating into muscle tissue, a whole set of genes for specific muscle proteins is turned on. The same genes remain silent in other kinds of tissues. Current work on these and other systems is concentrating on the regulatory sequences that flank the genes in DNA. Are there special DNA signals preceding muscle genes? What biochemical events trigger those signals to permit gene expression at the right time and in the right place?

In keeping with the great flexibility of DNA in supplying the needs of organisms, gene expression also is modulated by very different means. In *Drosophila* and *Xenopus*, some gene products are required at very high levels at certain developmental stages. Their supply is amplified, not by increasing rates of transcription, but by a transient amplification of the number of copies of the genes.

One of the most exciting current efforts is beginning to unravel the relation between specific genes and the anatomical features of *Drosophila*. Classical genetic analysis identified sets of genes that control the normal development of wings and legs on the fly's thorax as well as special structures on the abdomen. Mutations in these genes result in peculiar flies. Sometimes, normal appendages are missing; sometimes, one appendage appears in the wrong place; sometimes, appendages are oddly shaped. Now we know that these gene sets are clustered on one *Drosophila* chromosome in a region at least 200,000 base pairs long.

Probing the relation of specific genes to anatomy. Shown above are, moving clockwise from the upper left, a normal male fruit fly, a four-winged fly, a fly with an antenna transformed into a leg, and an eight-legged fly. The pathways between the structure and actions of particular genes and bizarre anatomy are now being traced. [Edward B. Lewis, California Institute of Technology.]

Molecular cloning has yielded pure DNA segments corresponding to a large part of the region, and various genes have been located. These studies should help us to understand the way in which specific proteins direct interactions between cells and, ultimately, the formation of the characteristic anatomical features of animals and plants.

Aging can be thought of as part of the developmental continuum. Investigation of development at the molecular level has provoked many new hypotheses about aging. Actually, we do not know whether the key to aging lies in DNA or in other biochemical changes. If it lies in DNA, we must ask whether there is an intrinsic program directing the events associated with aging or whether it results from accumulated random changes in DNA. Is it possible, for example, that accumulated reassortments of DNA by recombination contribute to the deterioration of critical metabolic systems? Active investigation of those and other ideas can be expected in the next few years.

Outlook

Until the early part of the 1970's, the direct study of the structure of genes was essentially limited to the DNA of bacteria and viruses, and even that was restricted. The sheer complexity of the DNA of plants and animals, some millions of times larger than that of simple viruses, precluded more than a glance at its organization and a few hints at how it functioned. We knew that the vast genetic system encoded in DNA included the information stored in genes in the form of the genetic code. We also knew that other DNA segments were important in regulating the time and the place in the body where the genes yield up their information in the form of the proteins that account for the structure and metabolic capabilities of cells and organisms. But we were stymied. Then, with the development of recombinant DNA techniques and, contemporaneously, greatly improved methods for the manipulation of DNA molecules, the picture changed dramatically. As a result, we have learned a great deal about the DNA of plants and animals, including man. We have been surprised to learn how different these complex organisms are from bacteria. For one thing, the genes in complex organisms are interrupted by other extraneous segments of DNA. Somehow, before the gene can function, these interruptions must be removed and the separate pieces of the gene spliced together. For another, the DNA segments that turn genes on and off at particular times and in particular places in the body are very complex, often being composed of several modules that are not even close neighbors of the gene or of one another. And it now seems that genes themselves make up only a portion of the DNA. As much as 20, 50, or even 90 percent of the DNA in different organisms has no known function as yet, and perhaps none at all.

Thirty years ago, genetic experiments with corn hinted that some parts of the DNA might be in flux, with DNA segments capable of moving about from one place to another in these large molecules. Now, we know that this is a general phenomenon and, in some cases, the events have been studied at the molecular level. Yet, we do not know the biochemical mechanisms involved in the movement, nor do we understand the implications of these jumping DNA segments. That great biological process, the development of a full plant or a complicated animal from a single cell, is now thought of as a precise program in which different genes are turned on and off in orderly fashion at different times and in different kinds of cells. In the case of antibody formation, the developmental program includes precise rearrangements of the DNA sequences themselves.

All of this new knowledge raises more questions than ever. As usual, in science, the most important outcome of the new knowledge is the revelation of unsuspected levels of ignorance. The next five years are likely to see progress in understanding splicing, the significance of movable elements, and the signals that turn genes on and off in response to genetically programmed changes or altered cellular environments. These processes are of much more than academic interest. Increasingly, it appears that the proper regulation of gene activity is fundamental to healthy organisms, be they crops, important animal food sources, or human beings.

Increasingly, it also appears that rearrangements in DNA may prove to be a common element in the diverse causes of tumors, including chemical carcinogens and viruses. Single genes that cause a normal cell to turn into a tumor cell are being studied. Many of these are a normal part of human DNA. What regulatory changes account for the conversion of a normal gene to a tumor gene? We are coming to understand that many diseases often have heritable components—if not the diseases themselves then a propensity toward them, or an inability to defend the body against them.

The past decade has been marked by an extraordinary pace of accomplishment. The techniques are in place and are being improved constantly. Many of the questions that need probing are clear. The next five years will provide some of the answers and will surely reveal a new generation of questions. At the same time, the new knowledge will be turned toward imaginative solutions for long-standing problems. New and better chemicals for manufacturing will be produced, new and better drugs and vaccines will be manufactured, new and improved plants may be developed, and diagnosis of diseases will be improved. With understanding also may come improved treatments for diseases that have resisted past efforts.

BIBLIOGRAPHY

W. F. Anderson and E. G. Diacumakos. "Genetic Engineering in Mammalian Cells," *Scientific American*, Vol. 245, No. 1 (July 1981), pp. 106–121.

J. Michael Bishop. "Oncogenes," *Scientific American*, Vol. 246, No. 3 (March 1982), pp. 80–92.

P. Chambon. "Split Genes," *Scientific American*, Vol. 244, No. 5 (May 1981), pp. 60–71.

R. D. Kornberg and A. Klug. "The Nucleosome," *Scientific American*, Vol. 244, No. 2 (February 1981), pp. 52–64.

P. Leder. "The Genetics of Antibody Diversity," *Scientific American*, Vol. 246, No. 5 (May 1982), pp. 102–115.

S. E. Luria. *Life: The Unfinished Experiment*. New York: Charles Scribner's Sons, 1973.

P. B. Medawar and J. S. Medawar. *The Life Sciences: Current Ideas of Biology*. New York: Harper and Row, 1977.

P. W. J. Rigby. "The Oncogenic Circle Closes," *Nature*, Vol. 297, No. 5866 (June 10–16, 1982), pp. 451–453.

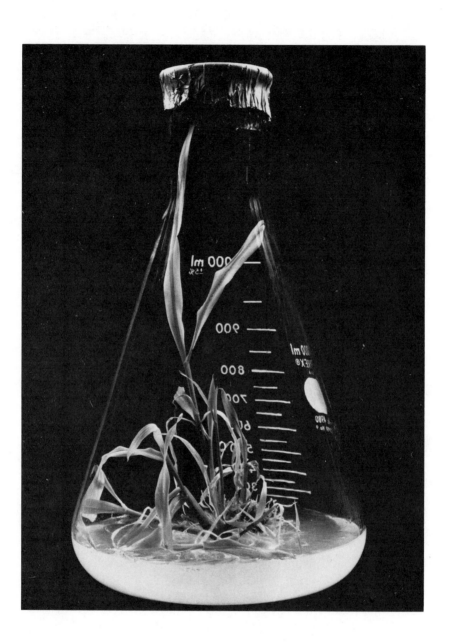

2

The Molecular and Genetic Technology of Plants

The recent and ready availability of tools and techniques for studying how genetic information is stored, expressed, shuffled, and passed from cell to cell has created an excitement among plant biologists that could not have been imagined five years ago. This excitement centers on the ever-increasing possibility that, as more details of plant genes become elucidated, new, faster, and less expensive ways of breeding crop plants suitable for changing demands and conditions will be discovered.

Foremost among the changing demands is increased production to support the growing population. It appears that some of the gains in production will have to be accomplished on marginal lands—lands that are too hot, too dry, too salty, or too polluted for our major crop plants. Successful use of marginal lands will require the adaptation of major crops to these harsh conditions and the adoption for cultivation of some of the underutilized plants, for example, grain amaranth and the winged bean. It is not at all certain that traditional crop-breeding practices can keep up with the rapidly changing demands on plants; after all, the great success of modern agriculture is based largely on crops that have been known and cultivated for thousands of years. Therefore, it is of great interest that the inquiry into the molecular details of plant genetics has reached a stage where application of the new knowledge and the new technology to help plants adapt to these demands seems within reach.

◀ A corn plant regenerated from tissue callus culture. [Molecular Genetics, Inc., Minnetonka, Minnesota.]

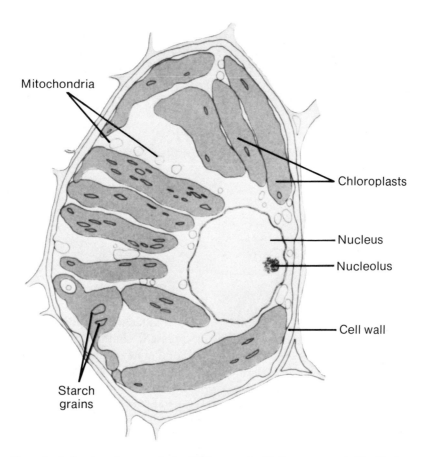

Figure 1   A drawing of a corn plant cell (*Zea mays*) with the components identified. This cell is typical of those surrounding the conducting tissues (xylem and phloem) of plant veins. Two of the organelles involved in energy production are visible—the chloroplasts and mitochondria. These bodies, along with the nucleus, contain the plant's genetic information. [SOURCE: Peter H. Raven et al. *Biology of Plants*. Third edition. New York: Worth Publishers, Inc., 1981, p. 15.]

To see how the new technology applies to plants, it is necessary to understand the arrangement of genetic information in plants.

THE PLANT GENES

General

A green plant carries its genes in three compartments—the nucleus, the mitochondrion, and the chloroplast. In general, the plant cell (figure 1) has one nucleus. This nucleus may carry one, two, four, or more copies of its genetic information. Thus, it is said to be haploid, diploid, tetraploid, etc. Each cell has from tens to hundreds of mitochondria and each mitochondrion carries several copies of the mitochondrial genome. Each cell has from one to tens of chloroplasts, and each chloroplast has several copies of its genome. For the proper functioning of a cell in division, growth, and development, the replication and expression of these three genomes must be coordinated. A brief description of each genome follows.

The Nuclear Genome   Intuitively, one might expect the amount of nuclear deoxyribosenucleic acid (DNA) per cell to be proportional to the complexity of the species in which the cell resides. But this is only roughly true. Animals, in general, have 1 to 5 picograms (pg) (10^{-12} grams) of DNA per haploid nucleus, but some salamanders have as much as 100 pg. Plants have up to 300 pg of DNA per haploid nucleus, but the range in the flowering plants is from less than 1 to 300 pg. In both plants and animals, only 1 to 10 percent of the total DNA is required to account for all known functions of DNA. Much of the total DNA consists of sequences tens to hundreds of base pairs long that are repeated hundreds to thousands of times and have no known functions. (See chapter 1 for a further discussion of this "excess" DNA and for an explanation of "base pairs.")

Most of the genes that code for specific proteins are present as a single copy or a very small number of copies. However, much of the single copy DNA is not transcribed and may have no function. Those single-copy or few-copy sequences of DNA that do code for proteins may be interrupted by noncoding sequences—introns. For example, the gene for at least one of the components of the leghemoglobin of soybeans, a protein in nitrogen-fixing root nodules (which is related to the ancestral globin gene of higher animals), has three intervening sequences. Two of the introns are located in almost exactly the same place as the two introns in the genes for the mammalian hemoglobin. The gene for the bean storage protein also has intervening sequences. In both cases the sequences correspond in several ways to analogous regions in mammalian genes. The challenge is to understand the way in which these intruding introns are cut out in transcribed ribosenucleic acid (RNA) molecules and how the freed ends rejoin to make a complete message for the synthesis of a protein.

The Mitochondrial Genome   Comparative studies of single-celled organisms, fungi, higher plants, and animals show that mitochondrial DNA is functionally conservative; that is, it encodes essentially the same genes in all organisms examined. These include the genes for the ribosomal and transfer RNA components of the mitochondrial protein synthesis system, three subunits of cytochrome oxidase, the apoprotein of cytochrome b, and one subunit of the adenosine triphosphate (ATP) synthase complex. These last three components are proteins central to cellular respiration. Although there is an absolute requirement for the expression of these genes in the biogenesis and functioning of mitochondria, most of the protein components of mitochondria are coded for by nuclear genes, synthesized in the cytoplasm, and assembled along with the components made in the mitochondria to complete the mitochondrial structure.

In view of the functional conservatism of all of the mitochondria examined, it was surprising to find that the structure of the mitochondrial genome varies greatly among the major groups. The mitochondrial genome of vertebrates and invertebrates is a 5-micrometer (μm) circumference of about 15,000 base pairs of double-stranded DNA, that of baker's yeast a 25-μm circle of 75,000 base pairs, and that of higher plants ranges up to 200-μm circles, or up to 650,000 base pairs. In addition, in higher plants, the DNA sequences are not the same in all of the circles. Therefore, the complexity, or the potential amount of information, is several times greater than would be expected if all circles were identical—as is the case for the yeast and animal mitochondrial genomes. It is important, then, to understand more about the structure and expression of the plant mitochondrial genome.

It has been found that the genome for yeast mitochondria contains intragenic and intergenic sequences that are not present in animal mitochondria. The coding segments are separated by noncoding sequences. Also, the number of introns in genes that code for some of the mitochondrial proteins varies from strain to strain of yeast. In contrast, mammalian mitochondrial DNA is a model of efficient information storage. The genes abut one another, with few or no intergene sequences. There are no introns.

In addition to this tremendous diversity among the major groups, there are differences—usually smaller—among closely related species which indicate that some features of the mitochondrial genome are changing more rapidly in evolutionary time than analogous features of the nuclear genome. As more and more details of the mitochondrial genome become available, there is renewed interest in examining these details for evidence of the postulated bacterial origin of the mitochondria. Strong evidence for such an origin comes from analysis of the base sequences of the small ribosomal RNA subunit of wheat mitochondria. A much greater similarity in sequence to the small ribosomal RNA of bacteria than could be expected by chance is shown. Furthermore, this similarity occurs in those sequences that are most conserved in the evolution of the bacterial ribosomal RNA. Determination of the complete sequence of the plant mitochondrial genome (the animal mitochondrial genome is already completely sequenced) would contribute greatly to ideas about the possible origins of mitochondria. However, because of the large size of the genome, this would be a monumental task.

The mitochondrial genome for maize is of special interest because in some lines it carries the information that causes cytoplasmic male sterility. In the past, these lines were a boon for the hybrid seed corn industry because they eliminated the necessity for expensive hand detasseling. However, one of the lines, the T (for Texas) male sterile strain, is susceptible to the toxin of *Helminthosporium maydis*—the Southern corn blight. An epidemic of this disease led to nearly disastrous losses of corn in several states in 1970 and 1971.

The Chloroplast Genome   In the chloroplast genome of higher plants, the DNA is double-stranded in a closed circle of 40 to 46 μm, about 135,000 to 150,000 base pairs. Although each chloroplast usually contains tens of DNA circles, every circle contains the same DNA sequence, in contrast to the complexity of the plant mitochondrial DNA. The chloroplast genome codes for perhaps 100 polypeptides. It also codes for the transfer RNA's, the ribosomal RNA's, and some of the elongation factors of the chloroplast protein synthesis system. Many of the proteins that make up the chloroplast are coded for in the nuclear genome. Thus, as in the case of mitochondria, assembly of a chloroplast requires considerable coordination in the expression of the nuclear and chloroplast genomes. For example, the small subunit of the photosynthetic carboxylase is coded for by the nucleus and synthesized in the cytoplasm. The large subunit is coded for by the chloroplast genome and synthesized in the chloroplast. The small subunits must be transported from the cytoplasm into the chloroplast, where eight of these small subunits associate with eight of the large subunits to make the functional carboxylase.

Maize and other C_4 plants are species that have evolved a special metabolism for the more efficient use of water and sunlight. In such plants, there are two distinct morphological types of chloroplasts—one in the bundle sheath cells and one in the mesophyll. Although the genomes of the two kinds of chloroplasts are identical, apparently the genes expressed are different in the two cell types. Also, the time of expression of these genes is determined by the developmental stage of the leaf. The metabolism of the C_4 plants has attracted much attention because of their high photosynthetic and water-use efficiency as compared with most crop plants. Similar attention is being given to control of the development of their chloroplasts.

As already mentioned, traditional breeding programs for crop improvement obviously have been successful and are, even at the present, our only certain way to further improvement. However, it now seems that we will soon learn just how plant gene expression is controlled. At that point, we can be cautiously optimistic about the possibility of using this knowledge to transform existing crop plants directly into plants that are more resistant to stress, more productive in marginal areas, more resistant to disease, and so forth.

TRANSFORMATION WITH *AGROBACTERIUM*

There is particular interest in the possibility of transforming higher plants by using one or more single gene traits that would alter the plant in a direction chosen by the investigator. One soil-borne bacterium, *Agrobacterium tumefaciens*, is already successful at transforming plant tissue. Crown gall disease arises when *Agrobacterium tumefaciens* injects some of its own DNA into the plant cells at wound sites. The resulting cancerous growth of the plant cells at the wound produces a gall. Such uncontrolled growth continues if a portion of the tumor is removed from the plant and placed on a simple nutrient medium. Most cells from normal or uninfected plants can grow on nutrient media only if they are supplied with the plant hormones auxin and cytokinin. The abnormal growth of crown gall tissue continues even after these bacteria in the excised tumors are killed. It was first suggested in the 1950's that this altered pattern of growth—transformation—was caused by a tumor-inducing principle that passed from the bacteria to the plant cells.

We now know that the tumor-inducing principle is a tiny segment of DNA containing several bacterial genes that do not appear to be expressed in the bacterium. Infectious strains of *Agrobacterium tumefaciens* contain large circular DNA molecules (see chapter 1 for a discussion of gene structure) called Ti-plasmids—tumor-inducing plasmids. These comprise 200,000 base pairs, enough information for about 100 genes. About one tenth of the DNA of the Ti-plasmid is transferred to the plant cell during the transformation process and incorporated into the plant's genetic information or genome, where it replicates during plant cell division along with the usual plant genes. This foreign bacterial DNA, T-DNA, also is transcribed along with the plant genes into messenger RNA, and the messenger RNA is translated into proteins. Thus, at least some of the bacterial DNA carried in the T-DNA fragment of the Ti-plasmid is expressed and is thought to be responsible for the pattern of uncontrolled growth.

Why has the crown gall bacterium developed this transformation system? The tumor cells produced by the system not only proliferate wildly, but they also synthesize amino acids not found in normal plant cells. These amino acid derivatives, "opines," are used as nutrients by the bacterium. In addition, the opines induce conjugation, or sexual recombination, of the bacteria. Thus, by introducing T-DNA into the plant nuclear genome, the bacteria induce the plant to produce not only rapidly growing cells that provide the space and the substrate for bacterial multiplication, but also unique chemicals that enhance the exchange of genetic information among the bacteria.

Clearly, it is of great importance to determine the exact sequences of those parts of the Ti-plasmid that are required for tumor induction. In that sequence lie the signals that enable the plant to recognize, integrate, and express the foreign DNA as well as signals that cause expression of selected plant genes. If it is possible to attach those signals to other genes of our choice, we should be able to get the plant to incorporate and express a gene related to improved quality, productivity, or pest resistance. Even without exact knowledge of those signals, DNA foreign to both the plant cell and the bacterium has been inserted into the T-DNA. This additional DNA is incorporated into the plant nuclear genome along with the T-DNA. Up to now, however, such DNA has not been expressed. Still, there is the exciting possibility that the T-DNA can be used as a general vector or vehicle to carry desirable genes into the plant.

TRANSFORMATION WITH OTHER VECTORS

Vectors other than the T-DNA of the Ti-plasmid of the crown gall bacterium are also of great interest. One good reason to seek out other vectors is that the host range of *Agrobacterium tumefaciens* does not extend to monocotyledenous plants such as grasses, lilies, irises, orchids, and palms. Therefore, no cereal grain can be transformed by *Agrobacterium*. However, as more is learned about how promoters control gene expression in plants, it may be possible to widen the host range of *Agrobacterium*.

What other vectors might be used to carry in new genes to bring about desirable transformations? Current hopes include the cauliflower mosaic virus, a small double-stranded DNA virus of about 8,000 base pairs that is easily cloned in bacteria and has been completely sequenced. This virus is taken up readily by plant cells of some species. In fact, naked DNA separated from the protein components of the virus particles can be used to infect plants simply by rubbing the DNA on some part of the plant; the virus particles then spread throughout the plant. The viral DNA replicates in the nucleus, probably as a plasmid. It then produces, via messenger RNA, viral structural proteins.

Bacterial DNA has been inserted into cauliflower mosaic virus, and these modified viruses have been used to infect plant cells. This bacterial DNA, which is foreign both to the virus and to the plant cells, persists in the virus throughout the viral replication cycle in the plant cells. From these results, it is clear that the cauliflower mosaic virus is a potential candidate for introducing desired DNA sequences into plant cells. To realize this potential, the part of the viral DNA that causes the disease in the plant would have to be removed and the desired DNA sequences inserted in its place. In addition, there are several other small, double-stranded DNA viruses that also could be tried as vectors if limitations are found in the carrier abilities of cauliflower mosaic virus.

Unfortunately, the host range of the cauliflower mosaic virus and of other small viruses is narrow, and vectors derived from these viruses might have only a limited possible application to a few crop plants. Also, movement of the viral DNA throughout the plant is possible only when the DNA is encapsulated in protein. There is little room in the protein capsid to add foreign DNA to the viral DNA.

In addition to other unevaluated DNA viruses, there are RNA viruses and viroids that might lend themselves as carriers. In some of the RNA viruses, a portion of the genome can replicate and spread in the plant without being encapsidated; thus, the size of the added foreign DNA might be unimportant.

Viroids (small, single-stranded, circular, unencapsidated RNA molecules) are of interest in their own right. They are the simplest pathogenic agents known. The potato spindle tuber virus, for example, contains only 359 bases. Viroids apparently do not code for protein, and it is not certain how they replicate or produce disease symptoms. Characteristics which make viroids of interest as possible vehicles are that they replicate in the nucleus, they can spread from the site of infection systemically and from cell to cell, and they are transmissible through the seed to the next generation.

The development of plant molecular vectors is still rudimentary, and our understanding of how plant transformation occurs is even more so. Nonetheless, we can get an idea of what lies ahead for higher plants by looking at what has been found in yeast. A leucine-requiring strain has been transformed by cloned homologous yeast DNA. The incoming transforming DNA integrates into the chromosomes and replaces the resident deficient gene. Exogenous yeast DNA also can transform yeast host cells without having the transforming DNA integrated into the yeast nuclear DNA. Autonomous, replicating DNA elements have been derived from yeast chromosomes. Selectable markers ligated, or attached, to these elements transform the host cells with high frequency while remaining outside the yeast chromosome. DNA sequences isolated from corn also will act as autonomous replicating sequences in yeast. Perhaps the most exciting behavior of these sequences is their ability to link stably with centromeric fragments derived from yeast DNA to form plasmids that are maintained without a selective pressure. These plasmids have been called "minichromosomes," and they clearly demonstrate that, at least in yeast, it is possible to "make" a new chromosome deliberately.

THE UPTAKE OF TRANSFORMING DNA BY PLANT CELLS

Once suitable vectors carrying the desired information are produced, they can be introduced directly into plant cells in the following way. Plant cells growing as a callus on a solid medium, in suspension on a liquid medium, or, in some cases, obtained directly from the leaves, are treated with fungal enzymes that degrade and remove the cell walls without damaging the living part of the cell, the protoplast.

Under appropriate conditions, these naked protoplasts will fuse with other plant protoplasts, with spheroplasts (bacterial cells with their walls removed), with lipid vesicles, with liposomes (small droplets of lipid material that can carry DNA within them), or with purified foreign DNA, such as the Ti-plasmid. By such fusions, the plant cell has introduced into it the entire genome of another plant cell, the entire genome of a bacterial cell, whatever kind of DNA (or RNA) one wishes to load into the liposome, or whatever kind of foreign DNA the protoplast will take up. Protoplasts derived from some plant species, notably, carrot, potato, tobacco, and petunia, will produce new walls around themselves and divide (1) to form an embryoid that will develop into a whole plant or (2) to form a callus from which roots and shoots will develop. A complete plant can be produced from these roots or shoots.

The capacity of a single cell or protoplast to regenerate into a whole plant is called totipotency. Such totipotency is retained by the cells of many tissues and organs of plants and is incontrovertible evidence that differentiation of embryonic cells into roots, stems, leaves, and other structures does not involve the loss of any of the information needed to make a whole plant.

In principle, a single plant protoplast can be transformed with foreign DNA and a transformed whole plant regenerated from that protoplast. In practice, fusions are induced at random among millions of plant cells and millions of the selected carrier agents. By this means, crown gall has been produced in tissue culture by infecting tobacco cell protoplasts with whole *Agrobacterium tumefaciens* cells, by infecting periwinkle protoplasts with *Agrobacterium tumefaciens* spheroplasts, or by allowing protoplasts to take up the Ti-plasmids. Calluses grown from these transformed cells developed stems which, when grafted onto a sturdy root stock, produced flowers. The stable transformants contained DNA sequences from the Ti-plasmid. Furthermore, when seeds were produced by crossing the transformed plants, the plants derived from the seeds, although not normal plants, still retained sequences from the Ti-plasmid.

THE REGENERATION OF WHOLE PLANTS FROM SINGLE PLANT CELLS

As indicated above, it may be possible to introduce into a whole plant new genes attached to a carrier that can move through the plant to transform existing cells as well as the reproductive cells that develop into the seed. For the present and the immediate future, however, it seems easier to attempt to transform plants at the cellular or isolated protoplast level. Although successful regeneration of a whole plant from a single cell or from a single protoplast is still more of an empirical art than a closely reasoned science, each year more species are added to the list of plants for which regeneration is possible. It is disappointing that there has been little success with some of our major crops, corn and soybeans, for example. However, other crop plants—potato, alfalfa, and carrot—can be regenerated from single somatic (nonreproductive) cells or from single protoplasts derived from somatic cells. In addition, cell cultures of tobacco, petunia, periwinkle, rose, and many other species not of primary interest with respect to food and feed production have been extremely accommodating and useful as experimental species for the development of the tools, techniques, and fundamental knowledge required to culture some of the less accommodating crop species.

The great German plant physiologist, G. Haberlandt, first tried to culture isolated plant tissues and organs as early as 1902. He failed, but correctly attributed his failure to not being able to supply the appropriate nutrient media. The idea that an isolated single cell could be totipotent and regenerate into a whole organism dates to Theodor Schwann in 1859. Following the discovery in 1928 of the plant growth hormone "auxin" (later identified as indoleacetic acid), several scientists in different countries in the 1930's and 1940's successfully cultured isolated plant organs and calluses derived from various plant tissues.

In particular, totipotency was achieved. In 1971, when whole tobacco plants were successfully regenerated from isolated protoplasts, the stage was set for today's advances.

One of the most successful and sustained studies of the regeneration of whole plants from protoplasts is the regeneration of potato plants from the protoplasts of potato leaf cells (see figure 2). Protoplasts were obtained from thin strips of leaves of Russet Burbank potatoes in solutions containing cell-wall-degrading enzymes from fungi. As the cell walls are degraded, the leaf cells first fall away from each other; as the

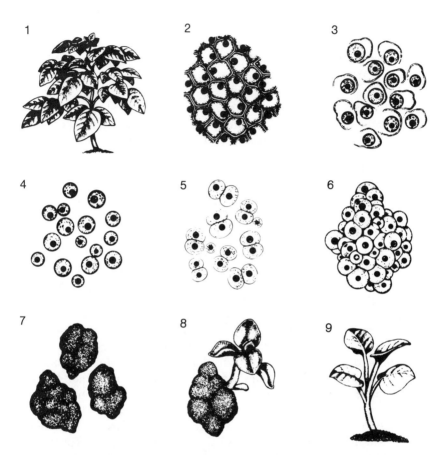

Figure 2 The cloning procedure employed to regenerate a complete potato plant from a leaf-cell protoplast is illustrated in this sequence of drawings. Small terminal leaves are first removed from a young potato plant (1). The leaves are placed in a solution containing a combination of enzymes capable of dissolving the cell wall (2). Another substance in the solution causes the protoplasts to withdraw from the cell wall and to become spherical, thereby protecting the living protoplasm during the disintegration of the wall (3). Next, the isolated protoplasts are transferred to a culture medium (4), where they grow, synthesize new cell walls, and begin to divide (5). After about two weeks' culture, each protoplast has given rise to a clump of undifferentiated cells, called a microcallus (6). The microcalluses are transferred to a second culture medium, where they develop into full-size calluses (7). At this stage, the cells of the callus begin to differentiate, forming a primordial shoot (8). The shoot develops into a small plant with roots in a third culture medium and is then planted in soil (9). [SOURCE: James F. Shepard. "The Regeneration of Potato Plants from Leaf-Cell Protoplasts." *Scientific American,* Vol. 246, No. 5 (May 1982), p. 156. Copyright © 1982 by Scientific American, Inc. All rights reserved.]

wall degradation continues, individual protoplasts are freed from their surrounding walls. After being removed from the cellular debris and the enzyme solution and placed in a suitable culture solution, some of these protoplasts regenerate walls and begin to divide, grow, and form calluses. After transfer to a different culture medium, the calluses form shoots. Finally, after transfer to still another culture medium, whole plants develop. Many of these plants are wildly aberrant, probably because they are derived from protoplasts that have somehow lost a part of their normal complement of chromosomes. Other variants retain most of the appearance and characteristics of the parent plant and the full complement of 48 chromosomes but, nonetheless, are observably different from each other and from the parent plant. These variants are of great interest because some of them appear to show increased resistance to the fungus that causes early blight, while others show resistance to the late blight fungus.

The plants derived from protoplasts also varied in tuber yield. Although none has shown a greater yield than the parent plant yet, the expression of yield variation, along with the reports of expression of possible increased disease resistance, tells us that quantitative plant traits can be modified by this technique.

The basis for the variation is not known. It is not likely to be due to point mutation—a change of a single base at a single place in the DNA structure—because the Russet Burbank potato is tetraploid and, therefore, each protoplast would have had four copies of each gene. Alteration of one of these copies would not be expected to have much effect. Also, the frequencies at which these variants appear are much greater than those of point mutations, even if the mutation were dominant or if the cells were haploid—having a single copy of each gene. There is a possibility that the variant property was present already, but not expressed, in the leaf cells from which the protoplasts were prepared, or that its expression was not observable because the expression was in a single cell only. Such differences in the somatic cells of a plant might occur by mutation or by recombination, an exchange of DNA between homologous chromosomes just before cell division.

Of course, it is also possible that the variants are somehow generated during the rather brutal removal of the walls to produce protoplasts, or that the culture conditions themselves may impose sufficient stress to produce the variants. Further, plants regenerated from cultured cells and protoplasts have many chromosomal rearrangements which may be associated with the mobilization of transposable elements. These are pieces of DNA that can move from one portion of the nuclear genome to another and, when inserted at a given locus, can control gene expression and cause large-scale rearrangement of adjacent DNA sequences. Regardless of how the variants arise, they offer a great opportunity to study further how the expression of genetic information is controlled as well as a long-range possibility of shortening greatly the time required to find agriculturally important variants.

THE NECESSITY FOR SELECTABLE MARKERS

The great strength of genetic techniques that use isolated protoplasts or suspensions of single cells is that one can deal with thousands to millions of individuals in one experiment. Because mutations, even when purposely brought about by mutagenic agents, are relatively rare events, it is helpful to be able to work with large numbers of individuals when one wishes to introduce variation by mutation. Even when variation is achieved by introducing a new gene through protoplast fusion with a liposome, only a small fraction (about one percent) of the protoplasts fuse successfully. In such experiments, one would like to use selectable markers that would, when the cells were placed in a suitable medium, permit the growth of only those cells that had acquired the new genetic information.

If one wishes to select mutants that are resistant to temperature extremes, extremes of pH, herbicides, toxins, heavy metals, pollutants, etc., that condition itself becomes the selective agent. Only the survivors are of interest. For other kinds of mutants, one must attach a selectable marker to the piece of DNA that also carries the desired gene plus whatever promoters are required.

One class of selectable markers is composed of bacterial genes that code for drug resistance. For example, Tn5, a bacterial transposon, a mobile piece of DNA conferring resistance to kanamycin in bacteria, has been used successfully as a selectable marker in yeast and in higher plant cell cultures as well.

Similarly, the bacterial transposon Tn7 codes for resistance to the drug methotrexate. It has been inserted into the T-DNA region of the Ti-plasmid and introduced into tobacco cells. The transformed cells acquire a resistance to methotrexate that the untransformed cells do not have. Thus, Tn7 may become a generally useful selectable marker.

The T-DNA region of the Ti-plasmid carries genes that can bring about auxin and cytokinin synthesis in the transformed cells. Therefore, the transformants can be selected by their ability to grow without added hormones. This selection procedure is of greatest value if the transformation frequency is 10^{-4} or greater, for false positives are occasionally seen. The octopine synthase gene in T-DNA can be used as a selectable marker for transformations at a frequency of $10^{-2}–10^{-3}$: it confers resistance to homoarginine, a toxic amino acid analogue.

Another possibility for a selectable marker lies in the use of cells that lack the nitrate reductase genes (at least two genes are required to bring about the synthesis of a functional nitrate reductase). Higher plants obtain the bulk of their required nitrogen from nitrates in the soil. Nitrate reductase catalyzes the first step in the assimilation of this nitrate. Cell lines lacking nitrate reductase can be recovered from cell cultures by standard tissue culture procedures. These cells fail to grow on nitrate. However, if amino acids are added to the culture medium, these same cells (or protoplasts) will grow into calluses and shoots. It seems almost certain that these cells could be transformed by adding the DNA of the missing nitrate gene (in a suitable vehicle). Thus, the nitrate reductase gene would be a powerful selectable marker. Only the transformed cells would grow on nitrate.

In bacterial and yeast genetics, many selectable markers have been generated because auxotrophs (cells that require specific substances for growth) for amino acids, vitamins, purine and pyrimidine bases, and so on were produced easily by mutation. As yet, such auxotrophs for higher plants are rare. However, a few examples of plants requiring the amino acid isoleucine were obtained by mutagenesis of haploid protoplasts. These plants had an absolute requirement for isoleucine, as did the calluses derived from those plants, and they had no detectable α-threonine deaminase—the first enzyme in isoleucine biosynthesis. Therefore, the plants would be an excellent source of protoplasts to be used as recipients in transformation studies using the α-threonine deaminase gene as a selectable marker. Production of these auxotrophs is one of the skills that must be developed before truly rapid progress can be made in applying molecular and genetic technology to a wide variety of agricultural problems.

THE ADVANTAGE OF HAPLOID CELLS AND PLANTS

The successful production and recovery of recessive mutants are greatly enhanced by the use of haploid cells rather than diploid (or higher ploidy) cells. Haploid cell lines have been produced from the pollen (usually from immature pollen cells still in the anthers of the flowers) of several members of the solanaceous species (petunia, tobacco, potato, etc.), and from gingko, belladonna, henbane, rye, barley, rape, certain genotypes of maize, and many other plants. For an immature pollen grain to develop into an embryo or into a callus, it must be diverted from its normal developmental path so that it will divide and begin to form an organized embryo directly or a disorganized callus competent to form shoots and roots. The procedures used to accomplish this regeneration are strictly empirical. Because of the importance of corn as a crop, there have been extensive efforts to regenerate maize by culturing tissue from the pollen sac. To date, the procedures are far from routine and reliable. When cell lines with the desired properties are found, the number of chromosomes per cell can be doubled by the use of colchicine to produce homozygous diploid cell lines; from these, homozygous plants are produced. Thus, fertile plants that breed true can be produced without the laborious and time-consuming process of backcrossing. The Chinese have had some success already in producing improved rice varieties derived from anther culture techniques.

As mentioned earlier, isolated plant protoplasts will fuse with each other, making it possible to bring together the genomes of plants that cannot be crossed by natural fertilization. The results of such a fusion are more complicated than is immediately apparent. The new cell produced by the fusion will have two nuclei that may or may not fuse. If nuclear fusion occurs, the chromosome number will be doubled. This may or may not be an advantage. If the protoplasts were derived from haploid cells, as for example from pollen culture, then the fusion nucleus would be diploid. A further complication of fusion between nonidentical protoplasts results because each plant cell has three genomes or three sets of genes: one in the nucleus, one in the chloroplasts (or in the plastids of a nongreen cell), and one in the mitochondria. The ten chloroplasts and hundreds of mitochondria per higher plant cell make it difficult to say how these nonnuclear genomes would sort themselves out in subsequent divisions of the fused cell. Nevertheless, protoplast fusion techniques may make it possible to cross sexually incompatible plants and thus bring together combinations of genes that cannot be brought together by traditional means. At the very least, fusion of nonidentical protoplasts will provide a new way to study the interactions of the nuclear genome with those of the mitochondrial and chloroplast genome—the cytoplasmic genomes.

Of course, these haploid cell lines also can be used to prepare haploid protoplasts to be recipients of DNA from other haploid cells by protoplast—protoplast fusion (parasexual hybrids) or to be the recipients of any DNA carried in by any suitable vehicle.

Implicit in all experiments with mutation and selection at the protoplast or cell level is the understanding that it will be possible to regenerate whole plants from the selected cells. Although this is nearly routine for such crop plants as potato and alfalfa and not too difficult for rice, it has not been possible to regenerate routinely whole plants from single cells or protoplasts of corn, wheat, and soybeans. Therefore, much emphasis is being placed on new techniques of embryoid genesis—the development from a single cell or from a small aggregate of cells of an embryolike structure that will grow into a mature plant. By empirical procedures, it has been found that such factors as the kind of nitrogen supplied—nitrate, ammonia, amino acids—and the relative concentrations of certain hormones, notably auxin, cytokinins, gibberellins, and abscissic acid, are of critical importance in embryoid genesis. However, a medium perfectly suited for the generation of embryoids from alfalfa cells, for example, is not necessarily suitable for any other species.

For many plants, including the cereal grains and some legumes, it is possible to develop routinely whole plants from callus cultures. Such cultures are derived by transferring a small group of cells (an explant) from some particular tissue to a nutrient medium. Explants taken from embryos or from growing regions—for example, the root apex, the shoot apex, the axial buds, and the intercalary meristems—grow readily as calluses in culture. For rice, sorghum, and maize, calluses derived from the scutellar (storage) tissue of immature embryos placed in culture seem to be the most useful. These calluses can be subcultured many times and still retain some embryogenetic or organogenetic capacity that permits regeneration of whole plants from them. Calluses derived from older tissues of the cereal grains cannot be regenerated easily into whole plants. It is obviously important to learn how to get this capacity expressed for all crop species in cultures derived from single cells. The capacity of a callus to produce cells that can regenerate into a whole plant is not only species dependent but also is very much dependent upon the particular genotype. Thus, there is an opportunity to compare readily and poorly regenerating genotypes to determine what factors are important in regeneration.

Outlook

Application of the new techniques of molecular genetics to plant science offers unique possibilities for manipulating plant genes to achieve the long-range objective of agricultural crop improvement. However, that first requires understanding plant gene regulation as it relates to gene structure. We currently understand little of the language in which instructions for gene regulation are encoded. Therefore, a first objective must be to approach this basic problem through the study of several important plant genes that will allow us to understand their regulatory signals. With this understanding, we can confront the increased demands for food and fiber resulting from population increases and improvements in living standards.

Some of the demand will have to be met by bringing marginal lands into production. The successful use of such land will require the development of crops that thrive under less than ideal conditions and the adaptation for cultivation of plants now underutilized, such as grain amaranth and the winged bean. Given this rising demand and that modern agriculture is based largely on crops known and cultivated thousands of years ago, the potential of new inquiries into plant genetics and development is enormously significant.

Indeed, the advances in genetic knowledge of the past few years promise ways of breeding new strains of plants that are faster and less expensive than traditional methods. Among these are transferring a gene trait via a soil-borne bacterium, *Agrobacterium tumefaciens*, or by using viruses and viroids as vectors to introduce desired genomes into a plant cell. The value of using such genetic carriers is that of transforming higher plants in a planned direction. However, for now, the technique has potential but not practice. The reasons lie in a still limited understanding of plant genetics. Thus, *Agrobacterium tumefaciens* injects its own DNA into plant cells at wound sites. The result is cancerous growth, producing a crown gall. What is not known is the structure of the genetic sequence carrying the sequence required for tumor induction, including the signals responsible for integrating and expressing that sequence in plant cells. Even without knowledge of these signals, DNA foreign both to the plant cell and to the bacterium has been inserted into plants, along with the tumor-inducing sequence. But such DNA has not been expressed. With further work surely will come the ability to use *Agrobacterium*, or other carriers, to insert desirable genes into plants.

The satisfactory culmination of these gene transfer manipulations requires the ability to regenerate whole plants from single and selected cells or from protoplasts; what is called totipotency. In principle, a single plant protoplast can be transformed with foreign DNA and a transformed whole plant then generated. However, as indicated above, deliberate transformation of selected cells is still more potential than practical. And the generation of whole plants from a single cell or protoplast is more of an empirical art than closely reasoned science. Moreover, while some plants can be generated from individual cells, such as potato, alfalfa, and carrot, others, such as corn, wheat, and soybeans, cannot. Why some plants are experimentally totipotent and others are not remains inexplicable.

However, enough is becoming known of the details of the plant genome that we can begin to understand how viruses and bacteria manipulate the control of gene expression in the plant for their own benefit. Enough is also known so that we can carry out some simple genetic manipulations. This manipulative ability will enable us to learn how the plant controls the expression of its genes in normal growth and development and in response to various kinds of injuries.

The refinement of these techniques and the rapidly increasing knowledge of how genes are transferred have created a great sense of excitement and new possibilities among plant biologists. An increased understanding of how plants control gene expression will help us to continue improving crop plants and to fit them to our needs more closely than ever before.

BIBLIOGRAPHY

T. O. Diener. "Viroids," *Scientific American*, Vol. 244, No. 1 (January 1981), pp. 66–73.

R. T. Fraley et al. "Liposome-Mediated Delivery of Tobacco Mosaic Virus RNA into Tobacco Protoplasts: A Sensitive Assay for Monitoring Liposome–Protoplast Interactions," *Proceedings of the National Academy of Sciences*, Vol. 79, No. 6 (March 1982), pp. 1859–1863.

Genome Organization and Expression in Plants. Edited by C. J. Leaver. New York: Plenum Publishing Corp., 1980.

Bruno Gronenborn et al. "Propagation of Foreign DNA in Plants Using Cauliflower Mosaic Virus as Vector," *Nature*, Vol. 294, No. 5843 (December 24–31, 1981), pp. 773–776.

S. H. Howell. "Plant Molecular Vehicles: Potential Vectors for Introducing Foreign DNA into Plants (Pathogenic Bacteria and Viruses, Genomes, Maize, *Zea Mays*)," *Annual Review of Plant Physiology*, Vol. 33 (1982), pp. 609–650.

F. A. Krens et al. "*In Vitro* Transformation of Plant Protoplasts with Ti-plasmid DNA," *Nature*, Vol. 296, No. 5852 (March 4, 1982), pp. 72–74.

Pal Maliga. "Isolation, Characterization and Utilization of Mutant Cell Lines in Plants," in *International Review of Cytology, Supplement 11A, Perspectives in Plant Cell and Tissue Culture.* Edited by I. K. Vasil. New York: Academic Press, Inc., 1980, pp. 225–249.

Jean L. Marx. "Ti Plasmids as Gene Carriers," *Science*, Vol. 216, No. 4552 (June 18, 1982), p. 1305.

Eugene W. Nester and Tsune Kosuge. "Plasmids Specifying Plant Hyperplasias," *Annual Review of Microbiology*, Vol. 35 (1981), pp. 531–565.

D. Pental et al. "Cultural Studies on Nitrate Reductase Deficient Nicotiana Tabacum Mutant Protoplasts," *Zeitschrift für Pflanzenphysiologie*, Vol. 105, No. 3 (February 1982), pp. 219–227.

James F. Shepard. "The Regeneration of Potato Plants from Leaf-Cell Protoplasts," *Scientific American*, Vol. 246, No. 5 (May 1982), pp. 155–166.

V. Sidorov et al. "Isoleucine-Requiring Nicotiana Plant Deficient in Threonine Deaminase," *Nature*, Vol. 294, No. 5836 (November 5–11, 1981), pp. 87–88.

Dan T. Stinchcomb et al. "Eukaryotic DNA Segments Capable of Autonomous Replication in Yeast," *Proceedings of the National Academy of Sciences*, Vol. 77, No. 8 (August 1980), pp. 4559–4563.

Michael Tribe et al. *The Evaluation of Eukaryotic Cells.* Edward Arnold, 1982.

3

Cell Receptors
for Hormones
and Neurotransmitters

Around the turn of the century, scientists developed concepts of drug action which postulated that responsive cells contain complementary "receptor" molecules to which specific drugs would have to bind to exert their biological effects. According to this receptor concept, the presence of an appropriate receptor becomes the basis for a cell or organism to be influenced by a particular drug; conversely, no receptor—no response.

More recently, the concept of specific receptors has been extended from drugs to other types of signals to which certain cells in our bodies can respond; for example, the detection of light by the retina and of odors by the nose, the actions of hormones and chemicals that transmit nerve impulses, the cell regulatory influences of certain lipoproteins in blood plasma, and the discernment of antigens that elicit antibody production by cells of the immune system. In recent years, several types of receptors have been purified substantially and, in every instance, have turned out to be proteins that avidly bind their ligands (the substances with which they interact selectively to bring about their biological effects). Studies on receptors have yielded great insight into the molecular mechanisms by which many exogenous agents control cellular activities, and into the genesis of several diseases. Receptor research also has contributed to the discovery of certain biomolecules

---

◀ The neuromuscular junction of a frog magnified 4,000 times. The sinuous band is the terminal branch of a motor-neuron axon partially ensheathed by a Schwann cell. Its surroundings are muscle fiber. [Barbara Reese, National Institute of Neurological and Communicative Diseases and Stroke, National Institutes of Health.]

whose existence was previously unsuspected and to the design and evaluation of new synthetic drugs. This chapter considers current and future investigations on receptors and their likely applications solely in the context of the actions of mammalian hormones and neurotransmitters.

HORMONES AND NEUROTRANSMITTERS

By conveying information between cells, neurotransmitters and hormones play crucial roles in the harmonious coordination of the development and functions of the vast array of specialized tissues in higher animals.

A hormone is a chemical messenger secreted by specialized cells that regulates the metabolism, and in some instances also the development, of other types of cells. The cells that manufacture hormones secrete them into blood plasma or other extracellular fluids, by which routes they are distributed in the body, usually in an indiscriminate fashion. Some hormones, for example, insulin, thyroxin, and the steroid hormones secreted by the cortex (or outer zone) of the adrenal gland, influence a wide variety of cells. Other groups of hormones may affect a more restrictive set of target tissues. Catastrophic pathological consequences can ensue from a severe deficiency or from a surfeit of virtually any hormone.

Neurotransmitters are molecules used by most nerve cells (neurons) to transmit signals across specialized contact sites called synapses— the tiny gaps between the end of a nerve fiber and either another neuron or some other type of cell, such as that of skeletal or cardiac muscle. At first glance, neurotransmitters and hormones appear to be very different from one another. Neurotransmitters need diffuse over a distance of only 5×10^{-6} centimeters across synaptic space, while hormones usually are transported via the circulation, in which they travel an average distance of roughly 10 centimeters to reach their target cells. Neurotransmitter actions occur in thousandths of a second, in marked contrast to the minutes, hours, or even days that may elapse between the administration of a hormone and complete manifestation of all of its biological actions. Yet the distinction between a neurotransmitter and a hormone is a somewhat arbitrary one, since the basic function of both groups of substances is to transmit signals between cells. Moreover, certain neurotransmitter molecules also can act as hormones outside the nervous system in the same organism.

Many naturally occurring or synthetic drugs can either mimic (agonists) or prevent (antagonists) the responses of cells to hormones or neurotransmitters. Some of these substances have become valuable probes for receptor functions in addition to their important medical applications.

DISCOVERY AND PROPERTIES OF HORMONE AND NEUROTRANSMITTER RECEPTORS

Nothing was known about the chemical nature of hormone or neurotransmitter receptors until two decades ago, primarily because no methods were available to identify receptors in cell-free systems. Beginning about 1960, procedures became available to introduce radioisotopic atoms of extremely high specific radioactivity into various positions of endogenous hormone and neurotransmitter molecules, or their corresponding unnatural agonists or antagonists, with no accompanying loss of their biological activities. With the aid of these radioactively tagged substances, it was possible to investigate their binding to isolated cells or subcellular components.

The process of hunting for neurotransmitter and hormone receptors using labeled agonists or antagonists as probes is vulnerable to many experimental pitfalls; a major one is that the ligands often bind nonspecifically to many other macromolecules in tissues besides their receptors. Nevertheless, the approach followed has been remarkably fruitful. Using specific binding assays in conjunction with advanced techniques for purifying proteins of very low abundance in cells, investigators recently have succeeded in obtaining highly enriched or, in a few cases, apparently pure preparations of various receptors. These have been used for further characterization. There is now consensus that receptors have the following properties in common.

Receptors Are Proteins

All known hormone and neurotransmitter receptors are proteins which, in some instances, have specific carbohydrate or lipid substances attached to them. In responsive cells, receptors are present at very low levels, often less than 20,000 receptor molecules per cell. Receptors that have been purified extensively have molecular weights that range from 70,000 to 250,000, and some of the larger receptors contain more than one type of protein subunit. Receptors for all neurotransmitters, and for all hormones that are polypeptides or catecholamines, are embedded predominantly in cell membranes, with their ligand-binding sites exposed to the outer surfaces of target cells. This suggests that hormones and neurotransmitters that interact with these sorts of membrane receptors need not penetrate responsive cells to generate at least their short-term effects. By contrast, receptors for all steroid hormones are present in both the soluble part of the cytoplasm and in nuclei, and the steroid–receptor complexes shuttle between these two cellular compartments. Thyroid hormone receptors are found mainly in nuclei, but they also are associated with other cell

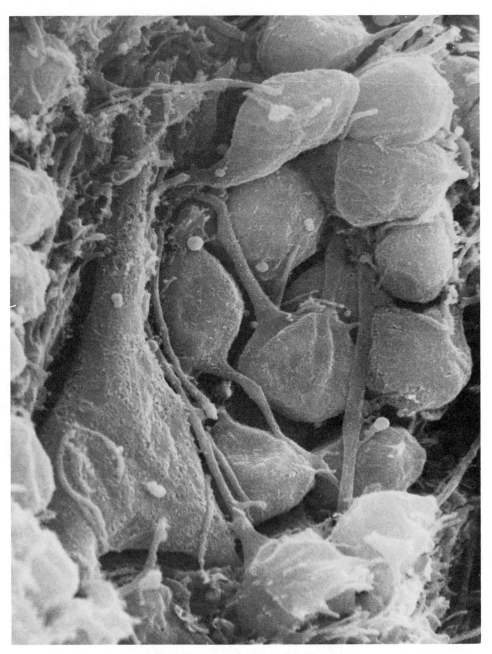

Scanning electron micrograph of granule cells of cerebellum. These small neurons obtain information from the stalklike dendritic processors reaching out to synapses, where signals are transferred by neurotransmitters via cellular receptors. [Barbara Reese, National Institute of Neurological and Communicative Diseases and Stroke, National Institutes of Health.]

structures. Steroid and thyroid hormones act inside cells, where they regulate gene expression processes.

The Two Functions of Receptors

Receptors have dual functions. One is to recognize and bind a particular type of hormone or neurotransmitter out of a sea of countless other molecules that encroach upon cells. The other is to transduce the informational signal inherent in the ligand–receptor complex in ways that alter the rates of preexisting biochemical reactions in the target cell. In other words, it is the receptor in union with a specific neurotransmitter or hormone—rather than the free chemical messenger—that serves as the actual regulatory material. It is hypothesized that, when their binding sites are occupied by appropriate agonists, receptor proteins undergo changes in conformation in their overall molecular shape, imparting to the receptor complexes the capacity to influence the responding cell's biochemical machinery. Hormone and neurotransmitter antagonists are believed to bind to the same receptor sites as the corresponding agonist molecules but in such a nonproductive fashion that the antagonist–receptor complexes are incapable of initiating biological effects. The exquisitely specific binding of hormones, neurotransmitters, and their antagonists to their receptors is usually extremely tight but, nonetheless, reversible.

Signal Amplification

Numerous intracellular biochemical mechanisms that amplify signals generated by receptor complexes have been described and, doubtless, many more remain to be elucidated. Most receptors are not enzymes—that is, they do not themselves catalyze any of the reactions of cell metabolism. In the case of many, but not all, receptors whose binding sites are exposed to the exterior of cells, occupancy of the binding sites by agonist ligands enables the receptor complex to influence an adjacent membrane regulatory protein. This protein stimulates a closely associated enzyme (adenylate cyclase), operating on the inner surface of the membrane. That enzyme synthesizes the "second messenger" substance, cyclic AMP (cAMP; or 3', 5'-cyclic adenylic acid).

Subsequent accumulation of cAMP in the cytoplasm provokes cascades of enzymatic transfers of phosphate groups to selected proteins (including other enzymes) whose functions are altered by such phosphorylation in ways that determine the ultimate biological responses. The formation of receptor complexes for other types of hormones on cell surfaces is coupled with other intracellular amplificatory biochemical events, such as changes in the levels of calcium inside the cells or other enzymatic processes that are independent of fluctuations in cAMP.

In the case of various steroid–receptor complexes, a major cellular amplificatory process is the regulation of the transcription of ribosenucleic acid (RNA) copies of specific genes on the nuclear chromatin, that transcription eventually leading to production of new proteins in the cytoplasm. In many cases, the formation of neurotransmitter–receptor complexes results in the rapid opening or closing of ion channels in nerve cell membranes. In turn, this causes changes in the fluxes of sodium, potassium, or chloride ions, thereby either triggering or depressing electrical action potentials in nerve fibers.

Receptor Formation and Turnover in Health and Disease

From the foregoing, it follows that the regulation of cell function via a receptor mechanism necessitates not only the arrival at appropriate locations in the target cells of adequate amounts of a particular hormone or neurotransmitter, but also the presence of adequate numbers of unoccupied functional receptors in the proper cellular districts; integrity of the various postreceptor amplificatory reactions also is necessary. Since receptors are proteins, instructions for their biosynthesis are encoded in corresponding genes. Thus, the production and turnover of receptor molecules can be influenced by countless genetic and environmental factors that affect the expression of their genes.

In several disease states, the symptoms mimic those of lack of a given hormone in the face of normal circulating concentrations of the hormone. The basic lesion seems to be the result of either inadequate numbers of functional receptors in the target tissue, or of a receptor gene mutation that results in the production of abnormal receptor proteins with deranged hormone-binding characteristics or conformations that attenuate the capacity of the receptor complex to relay signals to the cell's biochemical machinery. Such receptor dysfunctions usually render the animal or patient insensitive to the effects of even high doses of the same type of hormone.

In certain other endocrine or neurological diseases in which hormone or neurotransmitter production is essentially unimpaired, the symptoms reflect the formation of autoantibodies to the corresponding receptor proteins. Depending on the circumstances, the receptor autoantibodies can work in two ways. Either they can prevent receptor ligand binding and transducer functions (for example, with respect to the neurotransmitter acetylcholine in myasthenia gravis, a degenerative neurological disease), or they otherwise can mimic the agonist actions of hormones that act through a particular receptor mechanism (such as in certain hyperthyroid conditions in which the symptoms are reminiscent of excessive output of the thyroid-stimulating hormone by the anterior pituitary gland). Also of medical value has been the development of tests, based on measurement of the content of appropriate receptors in certain types of cancers, that have proved to be reliable guides for the selection of endocrine modalities of cancer treatment.

Other Practical Applications of Receptor Research

Correlations between the chemical structure of ligands and binding characteristics in experiments using receptor preparations have contributed to the design of new drugs that are agonists or antagonists for various classes of neurotransmitter or hormone receptors. The potential activity of new compounds can be assessed from quantitative estimates of their ability to displace known radioactively labeled ligands from high-affinity receptor binding sites. Comparable procedures have been invaluable for the detection and, then, the purification of naturally occurring substances with opiate or other pharmacological actions. Despite all this, it must be underscored that, to date, very few receptors for hormones and neurotransmitters have been completely purified. In most instances, the primary structures (the amino acid sequence) of receptor proteins or their subunits, or the precise chemical configurations of the ligand binding sites, remain to be deciphered.

The rest of this chapter illustrates in greater depth some of the foregoing principles with respect to a few selected types of hormone and neurotransmitter receptors.

RECEPTORS FOR GONADAL AND ADRENOCORTICAL STEROID HORMONES

The steroid hormones emanating from the testis, ovary, and adrenal cortex are classified into five categories on the basis of their biological properties:

- *Estrogens* are secreted extensively by the ovary and, to a much lesser extent, by the testis; they control primarily the development of female reproductive tissues.

- *Progestins* are secreted by the corpus luteum formed in the ovary as a result of ovulation; they affect such organs as the uterus and mammary gland and are essential for the maintenance of pregnancy. Natural progestins are precursors for the biosynthesis of androgens and estrogens in the gonads, and of the various adrenocortical steroid hormones.

- *Androgens* stimulate the development and functions of male reproductive tissues. Testosterone is the principal powerful androgen secreted by the testis, and weaker androgens are put out by the adrenal cortex in both sexes. Androgens also are biosynthetic precursors of estrogens.

- *Glucocorticoids* of the adrenal cortex regulate various reactions of protein and carbohydrate metabolism and facilitate adjustments of the organism to chronic stresses.

- *Mineralocorticoids* from the adrenal cortex influence sodium and potassium ion fluxes across cell membranes and contribute to the regulation of water balance.

Separate types of receptors for each of these classes of steroid hormones are found inside responsive cells.

The following is a brief summary of currently accepted concepts of the role of receptors in estrogen action. Estrogens cross cell membranes by passive diffusion or perhaps by transport mechanisms in certain physiological situations. On arriving inside target cells, the estrogens combine, without being changed chemically, with unoccupied cytoplasmic receptor proteins. Binding of the hormone permits the receptor, probably as a result of a conformational change, to undergo an activation process that proceeds optimally at body temperature and apparently gives rise to a complex of two receptor molecules with attached estrogen. The activated estrogen receptor complex is then taken up by the nucleus, in which it becomes associated with the chromatin, consisting of nuclear deoxyribosenucleic acid (DNA) combined with histones and other proteins.

The "acceptor" substances on chromatin with which estrogen-receptors interact appear to be mainly certain nonhistone proteins. By unknown mechanisms, the activated receptor complexes influence the synthesis in nuclei of RNA molecules that are complementary to various regions of the DNA genetic material, and especially the production of selected messenger RNA's that code for the biosynthesis of corresponding proteins. The estrogen-regulated RNA's then emerge from the nucleus into the cytoplasm where, without hormonal intervention, they are utilized for the new synthesis of specific proteins whose accumulation in the cell determines the physiological responses.

Until a short time ago, our knowledge of the occurrence and distribution of receptors for estrogens and other steroid hormones was based exclusively on experiments using radioactively labeled hormones as markers for their receptor proteins. Recently, a novel approach to receptor detection that does not depend on the presence of bound ligands has been developed. It was made possible by an availability of monoclonal antibodies that very specifically recognize and bind to discrete amino acid sequences in a given type of receptor protein. Monoclonal antibodies have been produced that bind avidly with estrogen–receptor complexes and also with unoccupied estrogen receptors. These antireceptor antibodies have been exploited as the basis of several techniques for estimation on ultramicro scale of receptor protein levels in cells and their subcellular compartments. These procedures have many practical advantages for both laboratory and clinical research, and they may soon replace older methods for receptor detection that are based on the binding of radioactive hormones.

Specific progestin receptors are found in the cytoplasms of cells from uterus, oviducts, vagina, mammary gland, anterior pituitary, and other tissues. The progestin–receptor complexes are translocated to the nucleus, where they enhance the formation of certain messenger RNA's. An interesting feature of progestin receptors is that their numbers in the cytoplasm of mammalian uterine and chick oviductal cells are greatly increased by prior administration of estrogens. This estrogen induction of progestin receptors meets a requirement for estrogen priming of female reproductive tissues for full manifestation of responses to progestins.

Separate receptor proteins that bind glucocorticoid and mineralocorticoid hormones are found in many tissues. These receptors also translocate from cytoplasm to nucleus, where the receptor–hormone complexes influence gene-directed synthesis of specific RNA molecules.

The role of distinct receptors in mediating the biological actions of androgens is complicated. In some responsive cells, circulating testosterone is quickly converted to a related substance known as 5α-dihydrotestosterone (DHT) by the enzyme steroid 5α-reductase. In those organs such as the prostate gland and penis in which DHT production from testosterone is extensive, the cytoplasmic androgen receptor binds preferentially to DHT, undergoes activation, and translocates to the nuclear chromatin. In other androgen-sensitive tissues in which DHT formation is negligible (certain muscles, kidney), the androgen receptor in union with testosterone is retained in the nucleus. It appears that there exists only a single species of androgen receptor that binds DHT even more tightly than testosterone. In several bioassay systems, the potency of DHT is greater than that of testosterone.

Steroid Sex Hormone Receptors and Disease

Research on steroid hormone receptors has proved to be of considerable medical significance as well as basic scientific importance. Two examples are cited.

The first concerns the value of estrogen receptor determinations as guides to the treatment of cancers of the human female breast. These and a variety of other tumors are frequently "hormone-dependent" or "hormone-responsive" in the sense that their growth in the body is either diminished by maneuvers that deplete blood plasma of appropriate hormones or by hormone antagonists, or otherwise may be influenced by the administration of exogenous hormonally active substances.

About 30 percent of women with actively growing mammary carcinomas have substantial quantities of cytoplasmic estrogen receptors in their primary and metastatic tumor masses. The other 70 percent of patients are "receptor-poor," that is, estrogen–receptor levels in their cancer cells are either undetectable or fall beneath a very low value. Receptor-poor patients—those whose breast cancers would not be expected to be estrogen-responsive—hardly ever benefit from endocrinologic treatments such as surgical removal of the adrenal glands or the anterior pituitary, excision of the ovaries in premenopausal women, or treatment with estrogen antagonists. Yet, roughly two thirds of patients whose tumors are rich in estrogen receptors undergo at least temporary objective remissions following endocrine therapies. Thus, on the basis of estrogen receptor measurements, individuals with receptor-poor inoperable breast cancers may be selected for immediate nonendocrine chemotherapy and spared from useless surgical removal of endocrine organs or the administration of estrogen antagonists. High levels of both estrogen and progestin receptors in the same mammary carcinoma are said to represent an even better criterion for the selection of patients for endocrine treatments. Similarly, the possible value of androgen receptor determinations as markers of the well-known responsivity of many human prostate cancers to androgenic hormones is under investigation.

A second example of clinical interest concerns the insights that have been obtained from androgen–receptor studies into the genesis of certain rare birth defects known as male pseudohermaphroditism. Affected individuals have a normal male sex chromosome identity, the XY pair, and, also, testes instead of ovaries, yet their other bodily sex characteristics develop in a feminine direction. Such disorders can result from a number of endocrine dysfunctions, some of which relate to disturbances in androgen receptors.

The diseases included in androgen-insensitivity syndromes (also called testicular feminization) appear to result from defective androgen–receptor proteins. Our comprehension of these conditions is based on the mechanisms of the initial differentiation of male extragonadal reproductive organs during fetal life, which takes place after the formation of fetal testes. Relatively early in development, the mammalian fetus contains four structures that look alike in both male and female fetuses of the same age (about the eighth week of conception in the human). These are:

- The two Wolffian ducts (precursors of the epididymis, vas deferens, and seminal vesicles in males, and which involute during female fetal sex development).

- The two Müllerian ducts (which develop into oviducts, uterus, and upper vagina in females, but disappear during male sex differentiation).

- The urogenital sinus (from which the male prostate gland and the lower segment of the female vagina are derived).

- The external genital primordia (destined to become the penis and scrotum in males or the female vulva).

Normal fetal female sex development does not require any hormones from the fetal ovary or from the mother's circulation. By contrast, the fetal testis produces at critical stages of development two hormones that are mandatory for normal male sex differentiation. One of these fetal testicular hormones is a protein, known as the Müllerian duct-involuting hormone, which prevents the formation of oviducts, uterus, and upper vagina. The other hormone is testosterone, which is essential for the Wolffian ducts to be maintained and to differentiate into epididymis, vas deferens, and seminal vesicles via mechanisms that do not entail conversion of testosterone to DHT. Likewise, the formation of the penis, scrotum, and prostate gland are utterly dependent on fetal testicular testosterone which, in these instances, is largely changed into DHT in the fetal precursor tissues. These hormonal actions are summarized in figure 1. In line with all of this, the aforementioned male structures can be produced in female fetuses by the administration of testosterone or DHT.

Human patients with the so-called complete form of the androgen-insensitivity syndrome have testes in the abdomen or inguinal region, have no ovaries, are devoid of other internal reproductive organs of either sex (they lack epididymis, vas deferens, prostate, and seminal vesicles, and also any uterus, oviducts, or upper vagina), and present externally a typically female vulva leading into only a short vaginal pouch. At puberty, menstruation obviously does not commence, but the florid breast development of the normal adolescent girl is usually evident. In adulthood, these patients' testes secrete almost normal male levels of testosterone (which is readily converted in their labial skin fibroblasts into DHT) and excessive amounts of estrogens. These bizarre developmental mishaps are now beginning to be understood. Assume that the testis in fetal life produces a Müllerian duct-involuting hormone that prohibits differentiation of oviducts, uterus, and upper vagina, and also testosterone, but that all tissues are refractory to the androgen because of disturbances in the production and function of

Primitive urogenital system

B Definitive male

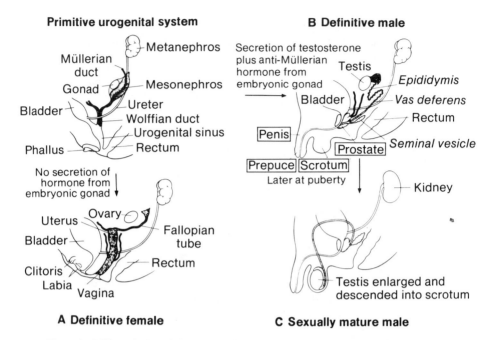

A Definitive female   **C Sexually mature male**

Figure 1 Differentiation of the mammalian genito-urinary tract in males and females. In the process of differentiation in the fetal male, structures requiring testosterone for differentiation are in italics; those requiring 5α-dihydrotestosterone are boxed. [SOURCE: W. I. P. Mainwaring. "The Androgens," in *Reproduction in Mammals: Book 7: Mechanisms of Hormone Action*. Edited by C. R. Austin and R. V. Short. Cambridge: Cambridge University Press, 1979, pp. 138–139.]

androgen receptor. The patients are incapable of responding not only to their endogenous testosterone, but also to this hormone or DHT given exogenously in huge doses. Three varieties of the androgen-insensitivity syndrome, involving various disorders of androgen receptors, have been characterized recently.

In another syndrome of male pseudohermaphroditism, characterized by deficiencies in the enzyme steroid 5α-reductase but with no attendant defects in androgen receptors, the patients exhibit feminized external genitalia and very small prostate glands but have well-developed epididymis, vas deferens, and seminal vesicle tissues. This clinical picture accords with the aforementioned requirement for a conversion of testosterone to DHT in appropriate precursor tissues at the time of embryonic male sex differentiation of the prostate and the external genitalia, but not in the Wolffian duct when it develops into the epididymis and seminal vesicle.

Thus, these very uncommon disorders of male sexual development have wider scientific implications. They provide telling evidence in favor of the theory, backed by extensive experimental studies in laboratory animals, that in all mammals, including man, there is a bias for the reproductive tract to differentiate during embryonic life in a female direction unless the fetal testis produces hormones that also are capable of acting on appropriate embryonic precursor tissues to induce the formation of characteristically male structures. This is true regardless of the sex chromosomal status of the individual. The concept has revolutionized our understanding of how development of the physical attributes of masculinity and femininity is determined in the fetus.

INSULIN RECEPTORS

Insulin is a protein hormone of molecular weight close to 6,000 that is secreted by the β-cells of pancreatic islet tissue. Insulin enhances the transport of glucose across the membranes of several types of cells, lowers blood sugar, and profoundly influences a host of cellular reactions involved in carbohydrate, fat, and protein metabolism. Highly specific receptors that bind insulin tenaciously are present on cell membranes and can be extracted from them by application of nonionic detergents. Solubilized insulin receptors have been purified more than 250,000-fold by methods that include adsorption of the hormone onto inert supporting materials to which insulin was attached chemically and which bind the hormone selectively; the insulin then can be removed by treatment with suitable reagents. There is a wealth of evidence that specific membrane receptors mediate the majority of the cellular effects of insulin, but the mechanisms by which insulin–receptor complexes on membrane surfaces transmit signals to the biochemical machinery of responsive cells remain nebulous, despite a plethora of hypotheses.

A severe lack or absence of insulin or, alternatively, the inability of insulin to work properly, has long been considered a cause of diabetes mellitus (hereafter referred to as diabetes). In its multiple clinical variants, this is by far the most common of the endocrine diseases, afflicting hundreds of millions of people. The symptoms of diabetes include abnormally high blood glucose levels (even in the fasting state), constant loss of large amounts of glucose in the urine, inadequate utilization of glucose by certain tissues, countless abnormalities of fat and protein metabolism, and often disastrous degeneration of blood vessels. In its severe forms, diabetes produces dehydration and coma, which, if untreated, kill the patients.

A minority of diabetics (10 to 20 percent) are able to secrete only tiny amounts of insulin because there are too few functional insulin-secreting cells in their pancreatic islets. These individuals require treatment with insulin to remain alive. A small number of diabetic patients harbor mutations that result either in defects in conversion of the precursor proinsulin molecule to insulin during biosynthesis of the hormone in the pancreas or, very rarely, in production of a structurally abnormal insulin in which a wrong amino acid becomes inserted at a defined position in the hormone molecule, with resultant attenuation of its biological effects. Many diabetics can secrete considerable amounts of normal insulin, although not in sufficient amounts to meet their metabolic demands, so that insulin treatment can be of ameliorative value in the individuals. In some of these patients, and in certain other disease states, derangements in insulin receptors seem to contribute to the symptomology.

Certain diabetic patients exhibiting elevated insulin concentrations in blood plasma and a diminished sensitivity to injected insulin often also have abnormally low numbers of insulin receptors in several tissues. The degree of insulin resistance is proportional to the decrease in insulin receptor levels; treatments that lower blood insulin and decrease the resistance to insulin tend to raise the tissue content of the receptors. Obese individuals who are not diabetics frequently have high blood insulin levels and a decrease in insulin receptor content of blood and fat cells that correlates with an insensitivity to insulin. This "down-regulation" of insulin receptors by insulin itself also can be demonstrated in cultured cells. It seems to involve the internalization of insulin–receptor complexes originally formed on the cell membrane with their eventual destruction inside the cell. In a rare disorder known as Type B extreme insulin resistance, associated with acanthosis nigricans (a skin lesion), a striking diminution in the binding of insulin to its receptor is sometimes associated with the appearance in blood plasma of antibodies that specifically interact with the receptor and block its ability to bind the hormone. The disease therefore appears to be an autoimmune condition in which the patients manufacture antibodies directed against their own insulin receptors.

CHOLINERGIC RECEPTORS

Acetylcholine, a fairly simple biomolecule, serves as a neurotransmitter at a variety of locations in the peripheral and central nervous system, and at motor endplates of skeletal muscles innervated by motor nerve fibers. There are two types of cholinergic receptors.

The one whose functions we know the most about is the nicotinic receptor, responsible for transmission at the nerve–muscle junction, and it is this receptor which is the subject of the following discussion. The nicotinic receptor binds acetylcholine at a site that also can accept the plant alkaloid nicotine, which acts like acetylcholine. The same receptor sites are antagonized by d-tubocurarine (the active ingredient

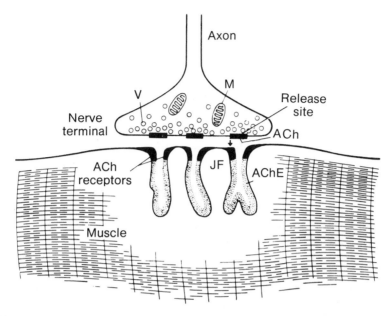

Figure 2 Diagram of a neuromuscular junction. Vesicles (V) discharge acetylcholine (ACh) at specialized release sites. After crossing the synaptic space (path indicated by arrow), ACh reaches the nicotinic cholinergic receptors, which are most densely situated at the peaks of the junctional folds (JF). Acetylcholine esterase (AChE) in the clefts rapidly hydrolyzes acetylcholine. M denotes mitochondria. [SOURCE: Daniel B. Drachman. "Myasthenia Gravis," *New England Journal of Medicine*, Vol. 298, No. 3 (January 19, 1978), p. 136.]

of curare, a South American arrow poison) and certain snake venom neurotoxins. Exceptionally rich sources of nicotinic cholinergic receptors are the electric organs (electroplaques) of such electric fish as *Torpedo* and electric eels like *Electrophorus* which, when excited by their cholinergic nerves, generate enormous voltages that can be lethal to other creatures.

Some basic features of neurotransmission by acetylcholine across synapses at neuromuscular junctions are diagramed in figure 2. Acetylcholine is synthesized from choline and acetyl-CoA by the enzyme choline acetyltransferase located in the cytoplasm of the presynaptic nerve ending. The neurotransmitter is stored in membrane-bounded vesicles in the nerve terminal, and each contains roughly 10,000 acetylcholine molecules. Transmission of the nerve impulse to the muscle involves release of acetylcholine from storage vesicles at specialized release sites into the synaptic space. After crossing the synaptic cleft, acetylcholine binds to the receptor that is most densely situated on the outer surfaces of the membranes at the peaks of the junctional folds.

When acetylcholine binds to the receptor, the latter undergoes a conformational change that produces a transient increase in permeability of the membrane to sodium and potassium ions, resulting in electrical depolarization of the membrane, which triggers muscular contraction. A similar series of events occurs in cholinergic neurotransmission across synapses between two nerve cells, with generation of a nerve impulse or action potential in the responding neuron. Acetylcholine esterase, the extremely active enzyme present in the synaptic cleft in neuronal synapses and the junctional folds of motor endplates in muscles, rapidly hydrolyzes much of the neurotransmitter to choline and acetic acid and thereby facilitates termination of the neurotransmission process.

Snake venom neurotoxins are small basic proteins that antagonize the binding of acetylcholine to nicotinic cholinergic receptors. These antagonists bind to the receptor very tightly. Radioisotopically labeled venom neurotoxins can be used to mark nicotinic cholinergic receptors in studies on receptor purification. After solubilization from fish electric organs by treatment with nonionic detergents, the receptor has been totally purified. This was accomplished by techniques that included the use of affinity chromatography on columns of inert supports to which neurotoxins had been attached, which specifically and selectively bind the receptor, and from which clean receptor can be recovered by elution with appropriate reagents. The nicotinic cholinergic receptor of *Torpedo* electric organs was shown to be an acidic glycoprotein complex, with a molecular weight of about 250,000, made up of four different types of glycopeptide chains, only one of which actually binds acetylcholine. The receptor molecule appears by electron microscopy as a rosette of subunits around a core, which presumably represents the ion channel. When acetylcholine attaches to the binding site, it induces a conformational change in the receptor molecule that results in increased transport of cations through the channel. One acetylcholine molecule opens a receptor channel that allows 10^4 sodium ions to cross the membrane in a millisecond. The channel is designed to remain open for only a very short time before the acetylcholine–receptor complex undergoes a second conformational change that closes the ion channel.

Research on antibodies directed against cholinergic nicotinic receptors has illuminated greatly our understanding of the disease myasthenia gravis. This condition, whose prevalence is about 1 in 20,000, is characterized by extensive muscle weakness, anatomic changes in the thymus gland, and increased production of a variety of autoantibodies and autoimmune symptoms. There is profound weakness and fatiguability of skeletal muscles, with a tendency for increased weakness with exercise and renewed strength with rest. Very frequently, the extraocular muscles of the eye are affected, resulting in ptosis—a drooping of the upper eyelids. But weakness of many other muscles often sets in, afflicting the arm and legs and causing poor function of chest muscles involved in breathing, which can result in death from respiratory failure.

Myasthenia gravis is essentially a disease of synaptic transmission at neuromuscular junctions. In 1973, it was observed that rabbits immunized with purified nicotinic cholinergic receptor developed symptoms that in many ways mimicked myasthenia gravis. Further research disclosed abnormalities in the postsynaptic membrane of neuromuscular junctions, as visualized by electron microscopy, and also an apparent decrease in the number of cholinergic receptors in neuromuscular junctions in patients with myasthenia gravis. It is now widely believed that the defects in neurotransmission and muscular weakness encountered in this disease are brought about by binding of autoantibodies to the receptor at neuromuscular junctions. Exactly how such autoantibodies exert their effects remains unanswered, although a resultant increase in the rate of cholinergic receptor degradation that exceeds the rate of receptor synthesis and thus produces a decrease in the net receptor levels appears likely; so does an attendant endplate membrane injury. Improvement of patients with myasthenia gravis by treatment with immunosuppressive drugs has been impressive, although short-term ameliorative therapy with inhibitors of acetylcholine esterases remains a standard form of treatment.

ADRENERGIC RECEPTORS IN RELATION TO CATECHOLAMINES AS HORMONES AND NEUROTRANSMITTERS

The term catecholamine is used widely to denote three naturally occurring substances that are closely related chemically: dopamine, norepinephrine (noradrenalin) and epinephrine (adrenaline). Norepinephrine and dopamine act as neurotransmitters in certain locations in the brain and the peripheral nervous system. Epinephrine and norepinephrine are released from sympathetic nerve endings into the general circulation, and from the adrenal medulla in response to countless acute stresses. Substantial amounts of dopamine are secreted into blood plasma from the adrenal medulla only in certain species. The catecholamines of adrenal medullary origin that enter the bloodstream exert hormonal actions on a wide variety of tissues. Dopamine produced by certain neurosecretory cells in the hypothalamus gets to the anterior pituitary gland via a local system of portal blood vessels and there serves as a hormone to depress the release of the protein hormone prolactin. A substantial proportion of the norepinephrine in peripheral blood plasma originates from sympathetic nerve endings that innervate blood vessels and other cells in many organs and the rest is derived from the adrenal medulla; in contrast, the latter tissue provides nearly all of the epinephrine in the circulation.

Catecholamines synthesized in cells in the adrenal medulla or sympathetic nerve endings are stored prior to their release in specific, membrane-bounded cytoplasmic granules, in conjunction with adenosine triphosphate (ATP) and special binding proteins. After their release, catecholamines can be removed by re-uptake by sympathetic nerve terminals. This is one process by which catecholamines have their action as neurotransmitters terminated; another is by metabolism of the catecholamines in certain tissues to pharmacologically inactive derivatives. Parenthetically, it may be noted that the re-uptake process is inhibited by certain so-called tricyclic antidepressant drugs such as desipramine as well as by amphetamine (both of which have medically important psychopharmacologic effects), and also by cocaine.

The pharmacodynamic and metabolic actions of catecholamines are legion. Recent research has permitted recognition of several classes of adrenergic receptors (denoted as α_1 and α_2, β_1 and β_2) and separate dopaminergic receptors. These adrenergic receptors are embedded in the membranes of responsive cells with their ligand binding sites exposed to the exterior. In some situations, both α and β receptors are present on the membrane of the same cell (for example, in fat, liver, and pancreatic islet β cells) in unequal proportions from a functional standpoint. When appropriate agonist substances bind to α- and β-adrenergic receptors in the same cell, they frequently elicit opposing physiological effects.

In many instances, occupancy by agonists of the binding sites on β-adrenergic receptors activates adenylate cyclase enzymes present in the same cell membrane, with resultant accumulation of cyclic AMP in the cytoplasm of the responding cell. This nucleotide then acts as a "second messenger" intermediate to elicit innumerable cellular biochemical events, which are entirely independent of the hormone-receptor complex and which underpin the biological responses. By contrast, α_1-adrenergic receptor-mediated effects are associated with a decrease in tissue cyclic AMP levels, whereas α_2 receptors do not appear to be coupled to adenylate cyclase action but, rather, to channels in the cell membrane that transport calcium.

Drugs that act selectively as agonists or antagonists of various types of adrenergic receptors have had therapeutic applications. Parkinson's disease, which usually begins in the elderly, is characterized by periodic tremors, disturbances of spontaneous and voluntary movements, and defects in posture. There is evidence that, in this disorder, there occurs increased cholinergic and decreased dopaminergic neurotransmissions in the basal ganglia, which lie beneath the cerebral cortex and close to the dorsal thalamus region of the brain. Attempts have been made to treat Parkinson's disease by the administration of anticholinergic drugs and especially by pharmacologic maneuvers aimed at increasing dopamine concentrations in the basal ganglia. Since dopamine itself, when given exogenously, does not get into the brain, the substance levodopa (L-dihydroxyphenylalanine—L-dopa), which does penetrate the brain and is converted into dopamine therein, has been used successfully in the therapy of Parkinsonism. Unfortunately, the treatment has unpleasant side effects, including nausea, vomiting, and disturbances of the heartbeat.

Certain antipsychotic drugs, such as chlorpromazine, are powerful antagonists of dopaminergic receptors, and this may account, at least in part, for the beneficial effects of these substances in certain types of psychosis. Propranolol, a drug that blocks the binding of catecholamines to both subclasses of β-adrenergic receptors and thereby influences neurotransmission at sympathetic nerve endings in blood vessels in the heart and elsewhere, has a valuable place in the treatment of certain types of hypertension. Bromocryptine, a chemically modified ergot alkaloid derivative, is a powerful agonist for dopamine receptors and can, as a result, depress the output of prolactin (the main stimulus to milk secretion by the mammary gland) by the anterior pituitary. Thus, bromocryptine can be used for the treatment of various disorders in which there is excessive secretion of prolactin into the general circulation. These disorders include decreased gonadal function and loss of libido in men and nonpregnant and nonlactating women associated with excessively high prolactin levels in blood plasma, and also galactorrhea, an excessive flow of milk that can occur in mothers who have ceased nursing and even in women without children.

LOOKING INTO THE FUTURE

Definitive research on receptors, which began only about 20 years ago, has expanded into an extraordinarily vigorous field. Yet our understanding of the structure and functions of receptors remains all too scanty. The dual demonstrations that receptors are bona fide molecular entities that represent the key components of the biochemical machinery by which cells respond to hormones and neurotransmitters, and that receptor malfunctions are implicated in the genesis of several diseases, have raised many more questions than they have answered. To close this overview with a mere litany of the many important unsolved problems would be of dubious value considering that prediction of progress in any branch of science is always hazardous and frequently

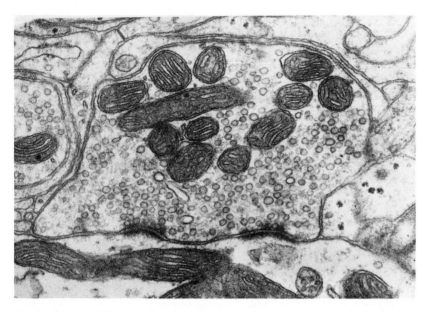

A synaptic terminal in the auditory brain stem. The cleft between post- and pre-synapses is in the lower part of the electron micrograph. A view of the active zone—where synaptic vesicles release their contents and receptors cluster on post-synaptic cells—is at the lower left. [R. Perkins and T. S. Reese, National Institute of Neurological and Communicative Diseases and Stroke, National Institutes of Health.]

fruitless. Nevertheless, a few vistas of future developments in receptor physiology merit brief consideration.

Until one can define in full detail the chemical architecture of the various receptor molecules involved in neurotransmitter and hormone action, it will be impossible to comprehend properly how the receptors recognize and bind their specific agonists or antagonists, or how the ligand–receptor complexes interact with those target cell components that transduce the regulatory signals to generate particular physiological responses. The exceedingly low concentration in which receptors are usually present in target cells has been a major stumbling block to isolating most receptors in a completely pure state and in sufficient quantities to enable their primary structures (amino acid sequences) and their conformational properties to be determined. However, it seems safe to predict that several other types of receptors besides the nicotinic cholinergic receptor will be purified completely in the near future. Thus, affinity chromatographic procedures, based on selective absorption of receptors to specific monoclonal antibodies attached to inert supports, promise a productive approach to receptor purification on a large scale. Another tantalizing way to obtain substantial amounts of pure receptors on their subunits would be the development of cultures of mutant cells that might overproduce certain types of receptors.

Now that recombinant DNA technology has progressed so tremendously, perhaps it will soon be possible to clone messenger ribosenucleic acids (mRNA's) that direct the biosynthesis of various neurotransmitter and hormone receptors. This might open the door to such advances as the synthesis of large amounts of mammalian receptor by bacteria, and to determination of the nucleotide sequences of the informational segments of cognate mRNA molecules from which the amino acid sequences of the corresponding receptor proteins could be deciphered. Pure receptor mRNA's also would be invaluable for probing into the structure of the corresponding DNA genes, which one suspects would contain a larger number of nucleotide sequences than those of the corresponding receptor mRNA molecules. Knowledge of receptor gene structure also would be very useful for further research on the mechanisms by which the production of specific receptor proteins are regulated in living cells.

Finally, there can be no doubt that studies utilizing isolated receptors have immense potential for increasing the success and cost effectiveness of the search for new therapeutic drugs of natural or synthetic origin. Prior to the advent of simple test tube procedures that can quickly determine the capacity of a particular substance to bind to a given type of receptor, the pharmacological activities of any new chemical suspected to act via receptor mechanism often had to be screened by observing responses in relatively large numbers of experimental animals or whole-organ preparations. For such purposes, it often has been necessary for chemists to prepare 25 grams or more of the compound which, like the biological assays, can be very tedious and expensive. By contrast, a large number of tests on isolated receptors usually can be performed with less than a milligram of the same chemical and in a much shorter time. Furthermore, if several substances were active in assays on intact animals, it was hard to judge whether one compound was more potent than another because of greater affinity for the receptor, decreased metabolism to inactive derivatives, or selective association with target organs. This made it difficult for researchers to design more potent and selective drugs.

Even though direct binding of a chemical to any particular receptor obviously does not necessarily vouchsafe that it will be pharmacologically effective in whole organisms, receptor screens surely will be applied on an increasingly wider scale to enhance the scientific vigor and diminish the expense of drug development. In this context, it is noteworthy that the enkephalins—a small group of polypeptides produced in the nervous system that regulate the mechanisms by which sensations of pain are generated—were discovered in studies on opiate receptors that recognize morphine and related substances from plants that act as pain killers. Further work on chemical modification of enkephalins as guided by opiate receptor studies promises the development of more potent and perhaps nonaddictive drugs for the control of intractable pain.

Outlook

Hormones and neurotransmitters coordinate the development and functions of the many specialized tissues of higher animal organisms by acting as chemical messengers that convey information between cells. These messengers bring about their biological effects by first combining with specific receptor molecules in responsive cells.

All known neurotransmitter and hormone receptors are proteins. It is the attachment of the messenger to a particular receptor protein that provides the actual signal to a responsive cell. In some cases, the same hormone may combine with more than one type of receptor and therefore can send cells different messages. If the appropriate receptor cell is lacking or defective, then the physiological responses to a hormone or neurotransmitter will be absent or attenuated.

Receptor studies have increasingly important medical applications. Testing the ability of new chemical substances to bind to specific receptors provides rapid and inexpensive screens for drugs that may potentially mimic or antagonize the actions of various hormones or neurotransmitters in living organisms. This may aid the development of therapeutically valuable drugs. Indeed, receptor studies contributed to the development of certain drugs of psychopharmacological importance, and of propranolol for the effective treatment of hypertension.

Since it is a combination of a messenger and its specific receptor that actually cues a cell's behavior, it is conceivable that certain diseases in which the symptoms resemble the result of too little or too much of a given hormone or neurotransmitter might actually reflect abnormal numbers or functions of a particular receptor, perhaps for genetic or other reasons. Certain rather uncommon forms of diabetes do, in fact, seem to be associated with defective insulin receptors rather than abnormally low levels of circulating insulin. Comparably, the neurological disease myasthenia gravis is apparently caused by the blockade of one class of neurotransmitter receptors by antibodies against these receptors that are manufactured by the patient's own immune system.

Genetically determined defects in receptors for male sex steroid hormones are associated with some rare disorders in which the patients have a normal male complement of sex chromosomes and develop testicles that readily produce testosterone, yet often develop other bodily sex characteristics that are female in appearance. Studies of these diseases, and of comparable disturbances of male sex development in animals, have revolutionized our understanding of the mechanisms by which male or female reproductive organs are formed in the embryo and develop further after birth and at puberty. Tests based on estrogen receptor determinations are of value in selecting alternative treatments for cancer of the human mammary gland. Research on receptors in relation to cancers responsive to several other types of hormones is now under way at many institutions in the United States and abroad.

Studies on hormone and neurotransmitter receptors in relation to the cause and treatment of human diseases are still very much in their infancy. There can be no doubt that this important new field will develop tremendously over the coming decade.

BIBLIOGRAPHY

J. D. Baxter and J. W. Funder. "Hormone Receptors," *New England Journal of Medicine*, Vol. 301, No. 21 (November 22, 1979), pp. 1149–1161.

J. R. Cooper et al. *The Biochemical Basis of Neuropharmacology*. Third edition. New York: Oxford University Press, 1978.

E. V. Jensen et al. "Hormone–Receptor Interaction in the Mechanism of Reproductive Hormone Action," in *Frontiers in Reproduction and Fertility Control*. Edited by R. O. Greep. Cambridge, Mass.: MIT Press, 1977, pp. 245–263.

W. I. P. Mainwaring. "The Androgens," in *Reproduction in Mammals: Book 7: Mechanisms of Hormone Action*. Edited by C. R. Austin and R. V. Short. Cambridge: Cambridge University Press, 1979, pp. 138–139.

J. Roth and S. I. Taylor. "Receptors for Peptide Hormones: Alterations in Diseases of Humans," *Annual Review of Physiology*, Vol. 44 (1982), pp. 639–651.

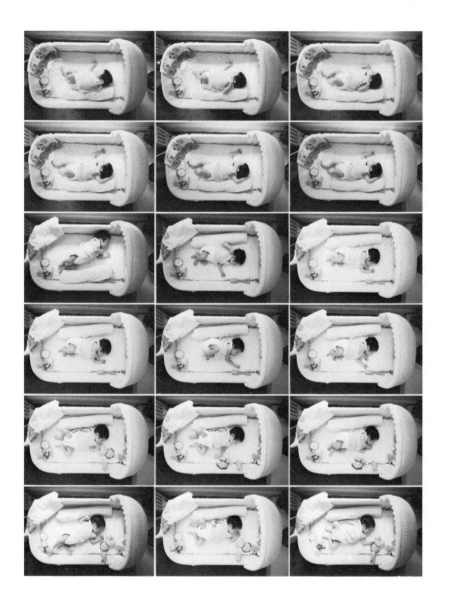

4

Psychobiology

Psychobiology is the scientific field that investigates the biological foundations of behavior and mental activity. It is a basic tenet of this research that all behavior originates in electrical and chemical events in the brain, which can be probed by the methods of modern biology. Similarly, such mental processes as perceptions, thoughts, and feelings are considered to be products of highly intricate but ultimately specifiable patterns of activity within the brain. The unraveling of these brain–behavior relationships is shedding light on fundamental questions about human nature and is leading to better ways of treating the serious neurological, communicative, and behavioral disorders that afflict a significant fraction of the world's population (see table 1).

Psychobiology is, of necessity, an interdisciplinary endeavor, since the multifaceted properties of the brain must be studied with a wide variety of techniques. Researchers with a background in chemistry investigate the chemical composition of the nerve cells in the brain and their secretions, anatomists examine the internal structures of the nervous system and their interconnections, while physiologists study the electrochemical processes by which nerve cells communicate. Psychologists and other behavioral scientists are charged with classifying and measuring the expressed behaviors and mental states and linking

◀ Photographs of sleepers taken at 15-minute intervals have shown that they move significantly about 10 to 20 times a night, that there are never more than 105 minutes without such movement, and that, typically, the longest quiescent periods are 60 to 75 minutes. [© Ted Spagna, 1983. All rights reserved.]

Table 1 Disorders of communicative processes and the nervous system

Condition	Number of people afflicted*
Deafness	2,000,000
Seriously impaired hearing	11,000,000
Blindness	500,000
Seriously impaired vision	1,400,000
Alcoholism and drug dependency	10,000,000
Convulsive disorders (e.g., epilepsy)	2,500,000
Stroke	2,500,000
Mental retardation due to congenital defects or birth injuries	2,500,000
Psychosis (e.g., schizophrenia)	2,500,000
Dementia (deterioration of mental ability)	2,000,000
Speech impairment (number of children in special school programs)	1,400,000
Aphasia (loss of language ability due to brain injury)	1,000,000
Head injury with resultant impairment	600,000
Parkinson's disease	500,000
Multiple sclerosis	200,000
Brain tumors	200,000

*Approximate numbers of people with these conditions in the United States, out of a total population of 220,000,000.

SOURCES: Mark R. Rosenzweig and Arnold L. Leiman. *Physiological Psychology*. Lexington, Mass.: D. C. Heath and Company, 1982, p. 4. From *National Research Strategy for Neurological and Communicative Disorders*, NIH Publication No. 79-1910, June 1, 1979.

them with specific events in the brain. Through concerted efforts on all of these fronts, researchers are beginning to glimpse some of the biological mechanisms responsible for the most refined of the brain's products—learning, thinking, conscious awareness, and language. This report considers recent achievements in these areas and prospects for future progress.

AN OVERVIEW OF THE NERVOUS SYSTEM

In vertebrates, the central nervous system (CNS) consists of the brain and the spinal cord, the latter mediating between the brain and the body. The human brain is a vast assemblage of some 50 billion individual cells called neurons, together with an even greater number of supporting cells known as neuroglia. The major subdivisions of the brain (see figure 1) include the cerebellum, which helps to coordinate bodily movements; the hypothalamus and pituitary gland, which regulate ba-

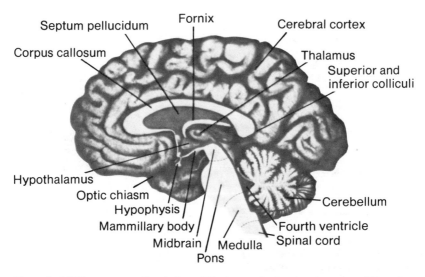

Septum pellucidum   Fornix   Cerebral cortex

Corpus callosum   Thalamus   Superior and inferior colliculi

Hypothalamus
Optic chiasm
Hypophysis
Mammillary body   Cerebellum
Midbrain   Medulla   Fourth ventricle
Pons   Spinal cord

Figure 1 Midline cross-sectional view of the human brain, showing some of the principal structural features. The medulla, pons, and midbrain are divisions of the brain stem. [SOURCE: E. R. Kandel and J. H. Schwartz. *Principles of Neural Science*. New York: Elsevier Science Publishing Co., Inc., 1981, p. 7. Copyright © 1981 by Elsevier Science Publishing Co., Inc. Reprinted by permission of the publisher.]

sic needs and instinctive behaviors; the thalamus, a relay center for sensory and motor information; and the paired cerebral hemispheres, covered with an extensive and deeply furrowed sheet of neurons some 2 millimeters thick, the cerebral cortex. The cortex shows a massive expansion in the more highly evolved mammals, particularly in the human brain, and contains much of the neural apparatus responsible for language and intellect.

Neurons

Neurons, the elementary units of the brain, are cells specialized for communication. Multiple extensions of the neuron's cell body, called dendrites, are designed to receive information from sensory cells or from other neurons. Also projecting from the cell body is a fiber that can be quite long (several feet in humans) called the axon, which transmits information to other neurons or to targets outside the CNS, including the muscles (see figure 2). By far the most connections made by neurons, however, are among themselves; a single neuron may receive inputs from hundreds or thousands of axons and send its own message to dozens of other neurons. From this point of view, the brain can be seen as an immensely vast and intricate network of interconnecting

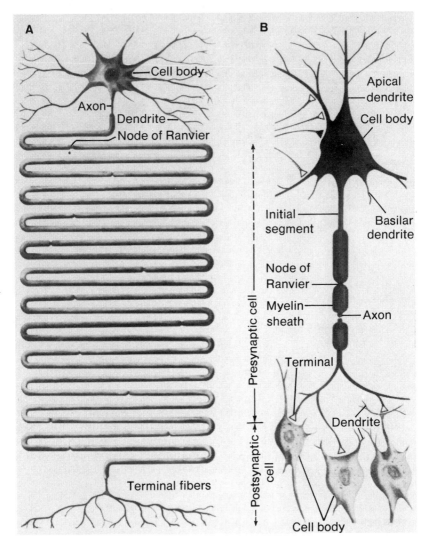

Figure 2 A schematic diagram of a typical neuron with its dendrites, axon, and cell body. The break in the axon signifies that it may extend for very long distances in relation to the size of the cell body. [SOURCE: E. R. Kandel and J. H. Schwartz. *Principles of Neural Science*. New York: Elsevier Science Publishing Co., Inc., 1981, p. 16. Copyright © 1981 by Elsevier Science Publishing Co., Inc. Reprinted by permission of the publisher.]

neurons. Recently developed neuroanatomical techniques that allow precise visualization of nerve cells and pathways are providing new insights into the ways neurons are assembled into larger structural units. In fact, using the new techniques of the neurosciences, more nerve circuits have been established during the past 10 years than have been defined in the entire history of brain research. Yet, this involves only a small fraction of the immensely complex structure of the brain.

The messages transmitted among neurons are in the form of brief surges of electrical current traveling along their outer membranes. Small currents in the dendrites and cell body may trigger larger action potentials that pass down the full length of the axon to influence the activity of target cells. The timing of action potential discharges represents a "code" that signifies how strongly a neuron has been affected by its inputs. As described below, neurophysiologists are attempting to decipher the more complex codes that represent sensory experiences, thoughts, and feelings by recording the patterns of these electrical potentials in the brain.

Neural Systems

The brain is organized into a number of systems of interconnected neural structures that are responsible for performing specific adaptive functions. Sensory systems transmit information about the environment or states of the body to specialized regions of the brain by way of sensory receptors in the eye, ear, tongue, skin, etc. Motor systems of the brain are responsible for producing accurate and timely movements of the body. Muscles are under the direct control of motor neurons in the spinal cord and brain stem, which in turn are controlled by specialized motor areas of the brain, including the cerebellum, basal ganglia, and motor cortex.

The limbic system is a set of brain structures and pathways that regulates motivational and emotional influences on behavior. Included in this system are the hippocampus, amygdala, and hypothalamus, all ancient structures observable in primitive vertebrates as well as in man. Closely affiliated with many limbic structures is the reticular activating system, a diffuse network of neurons in the brain stem and thalamus that regulates states of arousal and the sleeping/waking cycle.*

GROWTH AND DEVELOPMENT OF THE BRAIN

In the early stages of embryogenesis, the brain develops in all of its structural complexity from a very thin sheet of neurons. Under the influence of the genetic readout and through cell-to-cell interactions, these embryonic neurons proliferate, migrate to their proper places in

---

*A more detailed presentation of the basic organization of the brain can be found in the section on neuroscience in *Science and Technology: A Five-Year Outlook*, a report by the National Academy of Sciences published by W. H. Freeman and Company in 1979, and in the other sources listed in the bibliography.

the developing architecture of the brain, and make appropriate connections with other neurons or with target organs outside the CNS. One of the fundamental questions being asked about brain development is how neurons seek out the correct target cells with which to interconnect. The emerging view is that the outgrowth of axons and dendrites and the formation of new contacts between cells occur in a highly specific fashion rather than through random growth patterns. It has been shown, for example, that neurons in the retina of fishes grow into the visual sensory areas of the brain and create a point-to-point mapping of the eye upon these brain centers. The growing axons of retinal neurons must find their way over a considerable distance, bypassing numerous other neurons until they reach the cells that have the correct "identity" to make a permanent attachment. It is suspected that chemical codes guide this growth process, but their nature is unknown. A "nerve growth factor" has been identified that initiates the outgrowth of processes from certain classes of neurons. It is likely that the diffusion of such chemical factors may act over considerable distances to guide growing nerve fibers to their appropriate targets.

The developing nervous system is particularly vulnerable to such environmental insults as drug exposure, birth trauma, malnutrition, and infectious disease. Cerebral palsy and mental retardation are among the common disorders that can result from these agents. Through basic research on the neural and chemical factors that influence brain growth, it is hoped that new treatments will be devised to correct these and other developmental disorders. For example, it is estimated that the incidence of mental retardation can be reduced considerably over the next 20 years by applying the principles learned from research to genetic counseling, early diagnosis, improved nutrition, and appropriate environmental stimulation.

In the mature nervous system, cell multiplication and growth largely come to a standstill. The destruction of central neural tissue by disease or injury is not normally followed by an orderly regrowth of neurons to repair the damage, particularly in mammals. A long-term goal of research on developmental mechanisms is to discover methods of reactivating effective neuronal growth in the adult brain and for initiating the formation of functional connections to compensate for injury to the CNS. Recent experiments have shown that severing the normal input pathways to the hippocampus in the adult rat may lead to the growth of other neurons into the denervated regions and the formation of new connections. Continued research on neural "plasticity" in the adult brain should lead ultimately to ways of improving the recovery of function following brain damage.

Exciting new prospects have been opened by experiments showing that transplanted nervous tissue from the brains of young rodents will "take hold" and make connections with neurons in an adult animal's brain. These transplanted tissues also can serve as a bridge to guide the regrowth of severed axons. Current experiments with rats have attained success in restoring capabilities for learning and memory that were lost following lesions to the hippocampus by implanting fetal nerve cells at the site of the lesion. The possibility of replacing damaged brain areas and correcting associated chemical deficiencies (as in Parkinson's disease) by implanting appropriate neural tissue may thus be realized in the not too distant future.

NEUROTRANSMITTER SYSTEMS OF THE BRAIN

Neurons connect with one another at close-fitting junctions called synapses. In most cases, communication among neurons is mediated by chemicals called neurotransmitters. When an electrical impulse in the transmitting cell arrives at the synapse, tiny sacs (vesicles) of neurotransmitter are released through the cell's membrane into the junction and impinge upon specialized chemical receptors on the surface of the receiving cell. The transmitter induces electrical currents in the receiving cell that may enhance or suppress its generation of action potentials. Each neuron weighs the balance of these excitatory and inhibitory inputs and modulates its action potential discharges accordingly.

The neurotransmitters have been classified into several chemically distinctive groups. There appears to be nothing unique in the quantity or proportion of particular transmitters that sharply distinguishes the human brain from that of other mammals. In fact, the neurochemical systems encountered so far in mammalian brains and their general mechanisms of action seem to be quite similar to those seen in invertebrate nervous systems. The implication is that principles of nervous transmission derived from research on invertebrate nervous systems will have broad applicability to higher animals, including man.

For some neurotransmitter substances, the precursor chemicals as well as the intermediate stages of chemical synthesis have been identified. The chemical pathways by which "used" neurotransmitter is either inactivated or recycled for further use also have been worked out. Research on these mechanisms is yielding enormous dividends because most of the drugs used to treat psychiatric and neurological conditions have their impact on one or another phase of synaptic transmission.

Role in Mental Illness

Evidence is accumulating that different forms of mental disease, including schizophrenia and depression, are associated with abnormalities in specific neurotransmitter systems. Research on schizophrenia, the most severe and widespread of the psychotic diseases, has implicated an abnormality in the neurotransmitter dopamine. Antipsychotic drugs such as chlorpromazine, which are helpful in treating schizophrenia, reduce the action of dopamine in the brain, suggesting that this syndrome results from an overactivity in some portion of the dopamine system. A disorder called Gille de la Tourette syndrome, which produces grossly inappropriate speech and movement in children, also can be improved by drugs that reduce dopamine transmission. Affective disorders such as manic-depressive illnesses have been correlated with abnormalities in the activities of the transmitters norepinephrine and serotonin.

Drug therapies for these severe mental illnesses are being developed and improved continually as a result of research on the basic mechanisms of neurotransmission in the brain. Given the rate at which new information is being acquired on neurotransmitters and their actions (the list of naturally occurring brain chemicals that meet the scientific criteria of neurotransmitters has more than tripled in the past five years), continued progress is to be expected. However, better animal models of these mental disorders are urgently needed to facilitate further breakthroughs in their prevention, alleviation, and cure.

Role in Neurological Disease

Several types of serious neurological disorders have been linked with deficiencies of specific neurotransmitter systems.

Parkinson's disease, which produces abnormally slowed movements and tremors, results from the degeneration of neurons that use dopamine as a transmitter in certain structures of the basal ganglia of the brain. Great strides have been made in treating this disorder in recent years because of advances in our knowledge of the anatomical relationships of the key brain structures and the identification of neurotransmitter deficiency. It is now possible to treat Parkinsonism effectively by correcting the dopamine deficiency through oral administration of a precursor drug (L-dopa).

Huntington's chorea is a hereditary disorder that also involves the basal ganglia and manifests itself in uncontrollable jerky movements leading ultimately to mental deterioration and death. Recent experiments suggest that chorea may result from excessive dopamine activity in the basal ganglia (the converse of Parkinsonism) and, hence, its symptoms may be ameliorated by drugs that block dopaminergic neurotransmission.

Senile dementia is a condition of profound intellectual decline and personality change that occurs in some persons as they grow older. It is estimated that as many as 10 to 15 percent of persons over the age of 65 are affected, making this an increasingly serious problem as life expectancies in our society are extended. Studies of the brains of persons suffering from senile dementia and Alzheimer's disease (a type of dementia occurring in middle age) have revealed well-defined structural changes within the neurons and, in the past few years, marked decreases in the activity of the important central neurotransmitter acetylcholine have been documented. A substantial research effort is currently directed toward improving intellectual functions by compensating for this neurotransmitter deficiency, but the efficacy of this approach is still uncertain. Other mechanisms, including viruses and vascular diseases, also are under active investigation.

Role in Pain Reduction

One of the most important discoveries in recent years has been the identification of peptide neurotransmitters that occur naturally in the brain and have the same effect on the nervous system as the opiate drugs (for example, heroin and morphine) in alleviating pain and producing euphoric states. These endogenous, morphinelike substances (termed endorphins) have their impact at specific sites in the brain and spinal cord that modulate sensitivity to painful stimuli. Electrical stimulation of the brain at these opiate-related sites lowers the intensity of pain experiences, in part by interrupting the flow of pain messages from the spinal cord to the brain. There is evidence that the ancient procedure of acupuncture, the insertion and rotation of needles in specific locations of the body for blunting pain, may work by triggering the release of endorphins in the brain. This line of research into the neural basis of pain is likely to produce new therapies for individuals suffering from the chronic pain associated with many forms of disease and injury. While scientists are very optimistic about these discoveries in relation to pain, it has not been possible yet to establish how or whether the endorphin systems have anything directly to do with human opiate addiction.

Role in Behavior

Most neurotransmitter systems are widely distributed in the brain and probably serve a number of behavioral functions but, in some cases, there appear to be specific involvements. For example, an important functional system has been identified in a small nucleus of several thousand nerve cells in the brain stem region, the locus coeruleus. Neurons from this locus use norepinephrine as their transmitter and make connections with the full extent of the cerebral cortex, the sensory nuclei that relay information to the cortex, and many other brain regions as well. This small nucleus is thus in a position to exert a profound regulatory influence on the entire brain. Current evidence has implicated the locus coeruleus in the modulation of states of arousal and the sleeping/waking cycle, as well as in sensory information-processing. The sensitivity of neurons in the cortex can be enhanced by activity in the locus, making them more receptive to new sensory inputs. This could be an important mechanism for allowing the brain to pay attention to selected sources of information. Studies of this brain system may lead to techniques for correcting deficiencies of attention and learning that are seen, for example, in hyperactive children.

In some species of primates, the transmitter dopamine is particularly abundant in neurons of the frontal lobes of the cortex. Recently, it has been shown that the removal of dopamine from the frontal lobes by chemical techniques results in a severe learning disability in monkeys, making them unable to remember locations where rewards are being dispensed. The removal of other neurotransmitters does not seem to have any effect, suggesting that the dopamine neurons are essential for this type of learning and memory.

HORMONES

Many types of behavior, particularly expressions of aggression in sexuality, are strongly influenced by chemical substances (hormones) that are secreted into the general circulation from specialized glands. Much has been learned in recent years about the regulation of hormonal secretion by means of complex chemical messenger systems and about the mechanisms by which hormones influence behavior.

Specialized populations of target neurons have been identified in the brains of many vertebrates, generally within the limbic system, which respond to particular hormones and instigate the appropriate behavioral tendencies. One class of hormones consists of peptides that couple with receptor sites on the surfaces of target neurons in the manner of neurotransmitters; in contrast, the steroid hormones interact primarily with receptors inside the neurons to modify the cell's metabolic activities (see Chapter 3, Cell Receptors for Hormones and Neurotransmitters). Studies of these cellular mechanisms are leading to new concepts of how the genetic machinery of neurons is switched on and off by hormones to modulate complex behavior patterns.

In addition to their immediate effects on behavior, many hormones act on the developing nervous system to produce long-term alterations in brain structure and function and, ultimately, in behavioral traits. For example, adult sexual behavior in many mammalian species is determined in part by the impact of hormonal secretions on the brain near the time of birth. A significant amount of current research is aimed at discovering how hormones influence the early development of critical brain circuits and how they initiate elaborate behavior sequences through their actions on the target cells.

Another important research goal is to disentangle the contributions of heredity, learning, situational factors, and hormones to the expression of behavior. This work has obvious social implications for understanding the factors that affect reproductive behavior in humans, as well as the causes of aggression and violence.

NEURAL BASIS OF SENSATION AND PERCEPTION

A great deal has been learned in recent years about the brain's sensory reception systems for sight, hearing, touch, and pain and how these systems are organized to produce specific sensory perceptions. The principal method of investigation has been to record the activity of single neurons at different levels of the sensory pathways by means of tiny electrodes. The central finding has been that individual neurons are selectively responsive to a particular feature of a sensory stimulus, such as its color, location, shape, etc., or to a specific combination of features. This has led to the concept of "feature analysis" as a fundamental mechanism of sensory processing. In this analysis, a complex stimulus is "broken down" into its constituent features, which are encoded by individual neurons and then transmitted to higher levels of the brain. The cortical sensory areas play an essential role in the ultimate perception and recognition of sensory signals. For each of the sensory systems, there are multiple arrays of neurons that contain an orderly mapping of the spatial locations and other important features of the stimuli present in the environment.

Vision

The mechanisms by which the retina of the eye analyzes features have been delineated. The retina consists of a large sheet of light-sensitive receptor cells affixed to the rear of the eyeball, upon which the optic image of the world is focused, and two layers of nerve cells that analyze the information from the receptors. Connections between the color-sensitive receptors (cones) and neurons in the retina determine our capabilities for perceiving the normal range of color sensations. Color blindness results when one of the three light-detecting chemicals in the cones is either missing or not sufficiently distinct from the other two. The information about colors and contours in the world is transmitted from the retina to higher levels of the brain by way of the optic nerve.

Enkephalins, peptide transmitters, in a lobster's sight organ. Each spray, or rosette, is a cluster of light receptors. [Jorge Mancillas and Jacqueline F. McGinty, The Salk Institute.]

The role of the visual cortex in analyzing the information received from the retina has been demonstrated in an elegant fashion. Most neurons of the visual cortex are particularly sensitive to the straight lines within a visual scene; different columns of these neurons register lines of a particular slant and position. The sheet of neurons that makes up the primary visual cortex contains an orderly mapping of the visual fields, with inputs from the two eyes juxtaposed in adjacent "ocular dominance" columns. Some of these neurons are particularly responsive to moving lines, others to colored lines, and still others to lines at different distances from the eyes, giving rise to the perception of depth.

This research advances our understanding of the neural basis of perception, but many critical questions remain unanswered. In particular, we still need to find out how the individual features of a stimulus are "combined" to produce unified perceptions of whole objects and scenes and conscious awareness of the visual world. New conceptual approaches may help to answer this question; for example, it has been suggested that the visual system analyzes scenes according to their "spatial frequency" components (a mathematical way of representing its light/dark patterns) rather than simple stimulus features.

The orderly growth and development of the line-sensitive cells of the visual cortex may be permanently disrupted by abnormal sensory experience during critical periods of childhood. In experiments with cats, it was found that raising an animal in a visual environment that contained only, say, vertical lines resulted in a loss of function of the cortical cells sensitive to horizontal lines. An analogous effect is seen in children who grow up with severe optical defects (for example, astigmatism) that make lines at some orientations appear fuzzy. If their vision is not corrected early in life, they develop a permanent disability in detecting lines at those slants. The normal connections of the visual system also can be deranged if the two eyes are not aligned properly during the early childhood years, in which case the cells responsible for binocular depth perception may be lost. This research underlines the importance of early detection and correction of visual sensory disorders, and the same is true for other sensory disorders.

Innovative ways of compensating for some forms of blindness may emerge from basic research on the visual system. One approach to replacing the loss of sight is to stimulate the visual cortex electrically, thereby producing sensations of light that appear in a particular zone of the visual field. While these techniques can only create crude visual sensations at the present time, we may expect more refined visual experiences to be made possible as more is learned about the neural codes for visual perception.

Hearing

Research in the past 10 years has greatly increased knowledge of the peripheral auditory system and is leading to new prosthetic devices for overcoming deafness. The mechanisms by which the receptor cells (hair cells) of the inner ear are tuned to specific sound frequencies and how they transmit information to the brain by way of the auditory nerve have been worked out in considerable detail. Destructive changes in these delicate structures caused by such traumatic agents as intense sounds are being investigated with the aim of understanding the mechanisms of hearing loss.

To compensate for severe deafness resulting from inner ear damage, a new form of therapy is being investigated based on the principle of electrically stimulating the auditory nerve when it remains intact. In this procedure, a microphone picks up environmental sounds and controls the nerve stimulator in order to provoke auditory sensations in the brain. While the quality of hearing made possible by this technique remains poor, it does open a world of sound to patients long deprived of any auditory experience. Further advances in understanding the "codes" for specific sounds in the neurons of the auditory pathways (particularly for speech-related sounds) should improve the fidelity of hearing using this procedure.

Significant advances in the early diagnosis of hearing disorders have been made through computerized recordings of brain activity in the auditory pathways in man. The small brain potentials evoked by sounds in the auditory nerve and central pathway can be detected by electrodes placed on the scalp, thereby giving a direct measure of how effectively the peripheral auditory system is functioning. This test can reveal both the severity and type of hearing disorder and is particularly valuable in diagnosing hearing loss in infants before they begin to speak. Corrective measures such as hearing aids can then be initiated, before the development of speech is irrevocably retarded.

The Somatic Senses

Sensory information from the body (for touch, pain, position sense, etc.) is relayed through the spinal cord to the brain stem and ultimately to the cerebral cortex. Specific sets of small nerve cells and fibers in the spinal cord are designed to carry pain messages to the brain. The activity of these pain-transmitting neurons can be modulated or suppressed by stimulating other classes of nerve fibers, especially the larger ones that relay touch and vibratory sensations. These observations have led to the recent development of "neurostimulation" techniques for pain control; by applying electrical pulses or mechanical vibrations to specific zones of the body, chronic pain may be suppressed in some (but not all) patients. For example, if injury to a nerve in the hand resulted in severe chronic pain, electrical stimulation applied to the arm through a "cuff" may provide relief.

Further studies are needed to find out which patients are best aided by the pharmacological treatments for pain mentioned above and which will respond best to neurostimulation. Advances in the neurophysiology of pain mechanisms also should make it possible to deliver electrical stimuli in the most effective pattern and location to optimize pain relief.

LEARNING AND MEMORY

The successful adaptation of an animal to its environment or of a person to society is critically dependent upon learning. The ability to profit from experience is a characteristic of all animals, but the emergence of human language systems, cultural traditions, and technological devices is a consequence of the unique learning capabilities of our species. By the same token, individuals who are deficient in basic learning skills are especially handicapped in coping with the complexities of modern living.

Mechanisms of learning are being investigated by psychobiologists at a number of levels. Some evidence favors the view that learning involves alterations in the synaptic interconnections among neurons, but we need to find out much more about the underlying chemical and structural changes in individual nerve cells, the locations in the brain where such changes take place, and the mechanisms by which information is entered into and retrieved from the memory storage areas. The essential nature of the memory trace, whether it is localized to a few specific neurons or widely distributed in the brain, is still a mystery.

Elementary forms of learning have been investigated with considerable success in such relatively simple animals as worms, slugs, and insects, where the relevant nerve cells can be visualized and manipulated readily. These animals habituate to repeated stimuli in much the same manner as higher vertebrates do—that is, their initial startle or withdrawal response to a novel stimulus declines progressively as the once-startling event continues to recur. This modification of behavior results from a progressive decline in the amount of neurotransmitter released from the neurons registering the occurrence of the stimulus. In contrast, a change in behavior known as "sensitization" (in which the presentation of one stimulus makes the animal more reactive to another) occurs because of an increased release of transmitter from the same, modifiable synapse where habituation occurs. These synaptic mechanisms that mediate behavioral change may well apply on a wide scale, since there appear to be no essential differences in the structure or chemistry of synaptic transmission between simple invertebrates and higher mammals. A full understanding of the chemical changes responsible for learning may lead to the discovery of substances that can be administered to improve these functions.

The learning of new material takes place in at least two distinct phases in both animals and man. Newly acquired information is retained for the first few minutes in a temporary memory store that is vulnerable to disruption by such outside agents as a severe blow to the head. After several minutes to an hour or so, items that are well learned can be transferred to a more permanent form of memory that is less vulnerable to disruption and forgetting. Studies of patients who received electroconvulsive shock treatments as a psychiatric therapy indicate that memories become more and more firmly established, even over periods of years. This accords with the common observation that, as a person ages, remote events are at times remembered more vividly than recent happenings. Research along this line should lead to greater understanding of cognitive changes across the lifespan.

Several regions of the human brain have been implicated as playing critical roles in learning and memory storage. Studies of brain-damaged patients who have difficulty remembering newly learned material have shown that structures within the temporal lobe, notably the hippocampus, are important for the conversion of temporary memories into more permanent form. Patients who suffer damage to brain structures in the thalamus and hypothalamus as a result of chronic alcoholism also have difficulty forming strong memory traces and require extensive training before learning can occur. Interestingly, injury to either the temporal lobe or thalamic regions impairs the learning of specific facts to a greater extent than the learning of motor skills or problem-solving strategies. These studies of memory mechanisms are important for understanding the nature of mental retardation and senile dementia.

Cognition and Language

One of the greatest challenges facing psychobiologists is to understand the brain processes responsible for such higher cognitive activities as thought, language, and conscious awareness. Most current knowledge has come from studies of patients who have suffered injury to the brain or who have had neurological surgery. In recent years, the monitoring of ongoing brain activity through electrical recordings and mapping of blood flow patterns has added to the understanding of how different brain regions participate in mental activities.

Right–Left Asymmetries in Brain Function

A striking feature of the human brain is the specialization of the left and right cortical hemispheres for different cognitive, perceptual, and language functions. The left hemisphere (in most right-handed individuals) contains mechanisms for the expression and comprehension of language, for making logical deductions, and for mathematical skills. In contrast, the right hemisphere excels at visual perception, at judging spatial relationships, and at certain other nonverbal functions. This pattern of lateral specialization has been found in the vast majority of right-handed persons but, among the left-handed population, there is an increased proportion of persons having language functions in the right hemisphere or in both hemispheres. Accordingly, brain disease or injury that affects the right or left side of the brain is associated with predictable impairments in language and cognitive functions that depend in part on the patient's handedness. Recent experiments involving electrical stimulation of the language areas during brain surgery in alert patients have provided further insights into how the cortical areas of the left hemisphere are organized for the expression and comprehension of speech.

The different functions of the right and left hemispheres are illustrated dramatically in persons who have undergone, as a treatment for epilepsy, a surgical transection of the band of nerve fibers that interconnects the two hemispheres. This "split-brain" operation results in the emergence of two separate spheres of mental activity within the same brain. The isolated left side of the brain is capable of speaking fluently and of understanding spoken and written language, but is deficient at recognizing such complex visual stimuli as a person's face. The surgically separated right hemisphere, on the other hand, has a limited capacity for understanding language, but it succeeds admirably at the perception of spatial relationships.

Asymmetries and Anatomy   Investigations during the past 10 years have revealed that the pattern of cerebral specialization seen in man has definite neuroanatomical correlates. The association cortex of the temporal lobe that participates in the understanding of speech is expanded in the left hemisphere in most brains, while the frontal lobe is wider on the right. These brain asymmetries vary with a person's handedness. The temporal lobes of the great apes (chimpanzees, gorillas, orangutans) have left–right asymmetries less marked than those of man, while these asymmetries are completely lacking in several species of monkeys. Thus, from monkeys to apes to man, there appears to be a progressive increase in the degree of asymmetry of the brain areas that are critical for thought and language. There has been much debate about the evolutionary significance of lateral brain asymmetries, particularly in light of the highly asymmetrical brain organization for singing seen in certain species of birds.

In recent years, techniques have been developed for studying functional brain asymmetries in normal persons by presenting stimuli in the left or right sensory fields. These methods take advantage of the fact that sensory input to the right side of the body, the right ear, and the right half of the visual field is projected predominantly to the left cerebral hemisphere, and vice versa for left-sided stimuli. For example, it has been shown that, when spoken words are presented simultaneously in both ears, the word in the right ear tends to be recognized more accurately, presumably because of its preferential access to the language-dominant left hemisphere. Similarly, words flashed to the right side of the field of view can be identified faster than words on the left. The brain asymmetries revealed by these newly devised tests are providing a more complete picture of cognitive functioning in normal persons, and comparisons with individuals having sensory and intellectual impairment are under way in many laboratories.

Studies of lateral asymmetries in the human brain have important implications for helping persons who suffer from language disorders, learning disabilities, and other mental handicaps. An understanding of how the two hemispheres cooperate in producing these complex behaviors should lead to improved training techniques and educational environments for overcoming these deficiencies. Ultimately, this line of research should allow us to design more optimal educational experiences to bring forth the full intellectual capacities of each individual.

ORGANIZATION OF THE CEREBRAL CORTEX

The human brain is distinguished by a massive elaboration of the "association areas" of the cortex, so designated because they do not appear to have specific sensory or motor functions. It is commonly assumed that mankind's intellectual and linguistic capabilities derive in large measure from the activities of these cortical areas. The specific cognitive functions of the association areas of the parietal, frontal, and temporal lobes of the cortex are being clarified through studies of human patients with cortical damage and of monkey species that possess homologous brain areas.

In the parietal association cortex, sensory input converges from vision, hearing, and touch to form a spatial map of the environment and of the person's body within it. This region gives us the ability to orient ourselves with reference to external objects—for example, to read maps and find our way from one place to another, to visualize spatial relationships in three dimensions, and to link auditory and visual forms of information (as in reading). Within the past few years, it has been found that neurons within the parietal lobes of monkeys seem to play a special role in directing attention to events in the external world. Visual stimuli that capture the animal's attention activate neurons in this cortical area, while events that are ignored do not. Damage to these critical areas produces attentional deficits, such that significant objects in the environment may be neglected. This research is shedding considerable light on the mechanisms by which attention is focused upon important stimuli in the environment.

The apparent role of the temporal lobes and hippocampus in memory storage has been described already. In addition, association areas of the temporal lobe participate in the discrimination and recognition of complex auditory and visual stimuli. The enormous association areas of the frontal lobes allow a person to solve problems in a flexible and creative fashion and to make plans for the future. The ability to follow rules and adapt to society's constraints is also dependent upon frontal brain areas. We are still a long way, however, from understanding the neural mechanisms by which these intellectual functions are carried out in the association areas.

ELECTRICAL ACTIVITY OF THE BRAIN

As information is processed in the vast neural networks of the brain, the electrical potentials from millions of individual neurons are accumulated to form the "brain waves" of the electroencephalogram (EEG), recorded through the intact scalp. The rhythmic voltage oscillations of the EEG vary in frequency and amplitude according to a person's level of arousal, consciousness, and sleep. EEG recordings have demonstrated the existence of two distinct forms of sleep. In "slow wave sleep," the EEG is of high amplitude and low frequency, while in "rapid eye movement (REM) sleep," the EEG shows an increase in frequency, resembling the pattern seen in alert wakefulness. Studies during the 1960's revealed that most dreaming takes place during REM sleep.

While the biological functions of these two forms of sleep are still not well understood, a great deal has been learned about the brain regions and the neurotransmitter systems that control them. There is good evidence that slow wave sleep is initiated by the release of the transmitter serotonin from a small group of neurons in the brain stem, while in REM sleep, the locus coeruleus becomes virtually silent. Recent experiments have isolated chemical factors that promote sleep from the body fluids of both animals and humans. This finding may represent a breakthrough in the long search for a chemical "sleep substance" that builds up in the body to induce sleep. More research should lead to a better understanding of the numerous varieties of sleep disorders.

The waking EEG in man is characterized by the alpha rhythm (of about 10 cycles per second), which increases in amplitude during mental relaxation, particularly with the eyes closed. This "idling" rhythm is suppressed by a startle stimulus that requires attention or by mental effort applied to some task. The degree of alpha suppression in a particular brain region (for example, in one hemisphere) appears to be related to its level of information-processing activity.

Sensory stimuli elicit characteristic patterns of activity in both the sensory pathways and the cerebral cortex that can be detected in the form of small voltage changes in the scalp-recorded EEG. These "event-related potentials" (also called "evoked potentials") provide measures of how effectively sensory information is being delivered to the brain and of the perceptual and cognitive processes triggered by the stimulus. For example, when a person listens to speech, each word elicits a sequence of event-related potentials that signifies the transmission of the message from the inner ear to the cortex, the degree of attention that is paid to the words, and the person's understanding of their meaning.

New diagnostic tests of sensory and neurological disorders have been developed using event-related potentials recorded from the sensory pathways. By measuring the speed with which these potentials arrive at specific brain centers, it is possible to determine the extent to which the transmission of sensory information is impaired. The diagnosis of diseases that affect the sensory pathways, such as multiple sclerosis and peripheral nerve disease, has been facilitated greatly by these electrical techniques. In many cases, the objective evidence from the brain potentials is a more sensitive indicator of brain disease than are conventional neurological tests.

Some event-related potentials triggered from one to several tenths of a second after a stimulus seem to be markers of specific cognitive activities. An early negative potential is enlarged when a person's attention is directed toward a source of stimuli in the environment (such as a voice or a flashing light). Current studies of this brain wave component are aimed at understanding the causes and characteristics of attention-deficit disorders in children. Other types of event-related potentials are produced when a person recognizes a familiar stimulus, makes a decision, is surprised by a novel event, or has difficulty understanding a message. These potentials provide a continuous record of the timing of a person's mental acts and should lead to new insights into the structure of human cognition.

The recent development of highly sensitive magnetic detectors now makes it possible to measure the weak magnetic fields emitted by the human brain. The surges of minute electric currents in active populations of neurons induce magnetic fields to flow through the intact scalp, making it possible to localize those neurons more precisely than is possible with electrical recordings alone. This technique should aid in specifying which brain regions participate in various modes of sensory information-processing.

Mapping of Regional Brain Metabolism

Recent technical advances have made it possible to visualize the distribution of nerve cell activity throughout the human brain as different mental processes are carried out. By injecting an inert gas labeled with a radioactive tracer into the brain's blood supply, a two-dimensional mapping can be produced of those brain regions which are receiving higher levels of blood flow as a result of their increased metabolic and information-processing activity. Originally designed to examine cerebral blood flow in patients with brain disease, this procedure also can be used to chart the brain regions that are activated during sensory perception, voluntary movement, and more complex functions such as speaking, reading, and computation. Different mental activities have been found to be associated with characteristic patterns of regional brain metabolism within the cerebral hemispheres (see figure 3). Studies of schizophrenics have shown abnormal reductions in the metabolic activity of the frontal lobes, which may account in part for their disordered thought processes.

A more fine-grained mapping of the distribution of activity within the human brain has been achieved by injecting a radio-labeled sugar

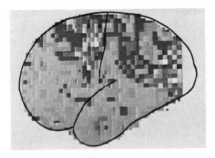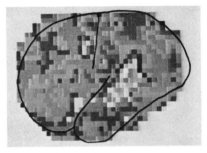

Figure 3 A graphic representation of blood flow through different regions of the cortex as a function of what the person is perceiving. At the left, the eyes are scanning a picture, resulting in greater blood flow (lighter shading) through the visual association cortex. At the right, the auditory cortical areas of the temporal lobe become active as the person listens to a spoken message. These patterns of cerebral metabolic activity were revealed by injecting a radioactive compound into the blood stream and measuring its distribution in the brain with an array of detectors outside the skull. [SOURCE: Niels A. Lassen et al. "Brain Function and Blood Flow," *Scientific American*, Vol. 239, No. 4 (October 1978), p. 63. Copyright © 1978 by Scientific American, Inc. All rights reserved.]

that resembles the brain's principal source of energy metabolism, glucose. The more active brain regions accumulate this tracer and emit radioactive particles (positrons) that can be detected at the scalp. This PET (positron emission tomography) scan allows the construction of a full, three-dimensional picture of the brain's ongoing activity. Using this method, it is possible, for example, to visualize the increased processing of information in the visual areas of the brain when a person is actively examining a pictorial scene; in contrast, patients with damage to the visual pathways show reductions of activity. This method is sensitive enough to distinguish the respective metabolic activities of the gray matter (containing the nerve cell bodies and their synaptic junctions) from those of the white matter (containing the insulated axons that connect distant nerve cells). PET scan measurements are made with a very low level of exposure to radioactivity and minimal risk to patients.

These new techniques for neurometabolic mapping show exceptional promise for revealing the brain areas that are malfunctioning in a variety of psychiatric and neurological diseases. Abnormal patterns of metabolic activity have been observed already in dementia of the Alzheimer's type, leaving little doubt that these techniques will be of value in the diagnosis and classification of age-related mental deficiencies.

In addition to their clinical utility, studies of regional cerebral metabolism are enhancing our knowledge of the enormously intricate patterning of brain activity that underlies normal perception and cognition. The combined use of electrical, magnetic, and metabolic mapping techniques will continue to sharpen the psychobiologist's concepts of how mental activities are organized in the brain.

Outlook

The fundamental scientific challenge for psychobiologists is to discover how the biological mechanisms of the brain give rise to complex adaptive behaviors and associated mental processes. While the relationship between mind and body remains, in many respects, as mysterious today as it was to the ancient philosophers, there has been an explosive growth in recent years of knowledge about how the brain is organized anatomically and how it functions at the biophysical, biochemical, and physiological levels to produce behavior. This research is producing new insights as to how the brain accomplishes complex mental acts such as learning, remembering, perceiving, and understanding, and how these functions become disturbed in mental illness. Indeed, the realization that mental and behavioral disorders can be linked to abnormalities in brain chemistry and physiology has led researchers to investigate biologically based therapies for a wide range of neurological, psychiatric, communicative, and cognitive disorders.

In the past few years, the principles that guide the development of the embryonic nervous system into the adult brain have been defined more precisely. Substances have been isolated that promote the growth of nerve cells and their interconnections. The transplanting of viable nervous tissue from one brain to another has been realized. It is likely that these discoveries will make it possible to compensate for many types of brain injury as well as for the developmental disorders that result in impaired behavioral functions. At the far end of the life span, specific changes in the structure and chemistry of nerve cells have been linked with the profound intellectual decline (dementia) that occurs in some persons as they grow older. While we are a long way from being able to intervene in the normal aging of the nervous system, there are good prospects for developing effective treatments for some forms of dementia over the next few years.

The chemical mechanisms by which nerve cells communicate are now understood in considerable detail, as are the modes of action of many neurotransmitter substances. This knowledge should result in more effective drug therapies for the many neurological and psychiatric disorders that are associated with abnormalities in specific neurotransmitter systems. The recent discovery of naturally occurring peptide neurotransmitters that exert the same effect on the brain as the opiate drugs, for example, morphine, has altered substantially current conceptions of the neural basis for pain perception. Research on these endorphin neurotransmitters and on the neurophysiology of the pathways that mediate pain is leading to improved methods for alleviating chronic pain associated with injury or disease.

The brain's ability to learn and remember underlies the development of both normal and abnormal behavior, since the majority of our intellectual and emotional characteristics are either learned or modified through life experience. Experiments on simple forms of learning in animals have revealed both the neural pathways involved and the cellular changes that make behavior modification possible. Investigations with human subjects have identified the brain structure critical for the formation of new classes of memories and learned skills. While there are still many gaps in our knowledge, these lines of research are converging toward a comprehensive understanding of how memories are stored and retrieved in the brain, and how they come to be expressed in thought and action.

New methods have been developed toward visualizing the intricate pattern of electrical and metabolic activities that forms a basis for higher cognitive functions. Complex mental acts such as speaking, reading, and paying attention have been associated with the characteristic patterns of energy metabolism and neural-electric activity in the brain that can be detected through the intact scalp. The combined use of these mapping techniques in conjunction with precise behavioral measurements should lead to new conceptual advances in understanding how mental processes are orchestrated in the myriad pathways of the brain.

BIBLIOGRAPHY

J. C. Eccles. *The Understanding of the Brain.* New York: McGraw-Hill, 1977.

David H. Hubel. "The Brain," *Scientific American*, Vol. 241, No. 3 (September 1979), pp. 44–53.

E. R. Kandel and J. H. Schwartz. *Principles of Neural Science.* New York: Elsevier Science Publishing Co., Inc., 1981.

National Academy of Sciences. *Science and Technology: A Five-Year Outlook.* San Francisco: W. H. Freeman and Company, 1979.

Mark R. Rosenzweig and Arnold L. Leiman. *Physiological Psychology.* Lexington, Mass.: D. C. Heath and Company, 1982.

5

Surface Science and Its Applications

Surface science deals with the few layers of atoms at the surfaces of solids and the marked effects these layers exert on the behavior of the solids themselves and on their interactions with gases, liquids, and other solids. Basic investigations in surface science have multiplied rapidly during the past 15 years; during this time, we have learned most of what we now know about solid surfaces at the atomic level. The vigorous efforts to acquire this knowledge have been notably sustained by the fundamental importance of surface phenomena to the many technologies that, for a variety of reasons, rely on solids.

Solid-state microelectronic devices, for example, are fabricated by the controlled growth of crystals at surfaces to form solid–solid junctions or interfaces having specific electronic or optical properties. The broad and important field of heterogeneous catalytic chemistry involves the interaction of gases with surfaces and the chemical transformation of molecules on surfaces. Technologically important solids increasingly must have large ratios of surface atoms to nonsurface or bulk atoms. Miniaturization of electronic devices means that a greater proportion of the atoms in a device lie on surfaces or at interfaces. Similarly, catalysts with large surface areas involve small, supported clusters of atoms as the active catalytic agents. The phenomena of corrosion and embrittlement of materials as well as such important

◀ The bright spots are images of individual atoms on the tip of an iridium crystal, seen through a field-ion microscope. [Bell Laboratories.]

technologies as nuclear energy, fossil fuel conversion, and solar energy extraction all present fundamental problems involving surfaces.

The tools of surface science have progressed remarkably. The chemical composition of surfaces can be determined with a sensitivity of 1 atom in 1,000 by techniques such as Auger electron spectroscopy (AES). Methods based on electron diffraction and the absorption and scattering of X-rays can determine distances between atoms at surfaces to better than 0.1 angstrom (0.01 nanometer). Tunable synchrotron radiation enables us to distinguish the chemical environment of a surface or interface atom from that of a bulk atom. A technique called molecular beam epitaxy, used to construct the crystal architecture of microelectronic devices, can deposit a layer of material 100 atoms thick that varies in thickness by only one or two atoms over the entire layer.

The outlook for improved understanding of surfaces and surface interaction phenomena is bright. Studies of chemical reactions at surfaces will be advanced by methods being developed to put interacting molecules from the gas phase into specific vibrational and rotational states and to determine these states as molecules leave the surface after interacting. Higher resolution methods involving infrared absorption and atom-beam scattering will improve greatly our ability to determine molecular conformations on surfaces. More intense and more readily available synchrotron light sources, in both the ultraviolet and X-ray regions of the electromagnetic spectrum, will refine our ability to probe both the geometrical and the electronic structures of surfaces and solid–solid interfaces.

THE NATURE OF SURFACE SCIENCE

Solids are one of the three traditional phases of matter—the others being liquids and gases—of which we and the physical universe about us are composed. All solids, whether natural or manmade, have surfaces. In the interaction of a solid with its environment, the action to a very large extent is at the surface of the solid. It is possible for X-rays and nuclear and cosmic radiations to interact importantly with solid matter without much regard to the surfaces. But the character of the surface of a solid can influence profoundly the electrical and chemical behavior of the interior of the solid and can be the determining factor in the interaction of the solid with electromagnetic radiation and the gaseous, liquid, and solid phases outside it.

The surface of a solid can be defined as the atoms that comprise the few, say three to five, outermost atomic layers of the solid. These layers differ appreciably in geometric and electronic structure from those in the bulk, the deep interior of the solid. The surface layers may

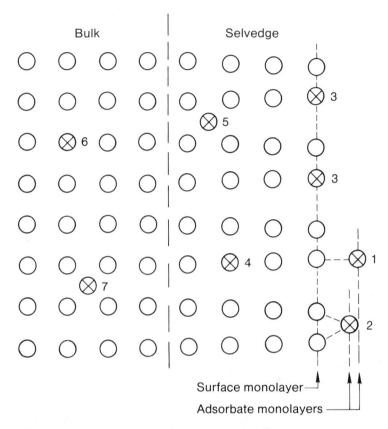

Figure 1 Schematic diagram showing a planar cut into the near surface region of a crystalline solid (the selvedge) that is affected crystallographically and electronically by the presence of the surface. Foreign atoms in various positions are identified by a cross in the circle representing the atom. Atoms in positions 1 and 2 are adsorbed to the surface in two modes of binding. Atoms in position 3 are absorbed or incorporated into the surface monolayer; atoms in positions 4 and 6 are substitutionally absorbed; those in positions 5 and 7 are interstitially absorbed into the selvedge and bulk, respectively.

consist only of atoms of the bulk solid, but they also may include foreign atoms that cling to, or are adsorbed at, the outermost layer, or they may occupy other specific positions in the surface region, which is sometimes called the selvedge (see figure 1).

The basic understanding of surfaces is really a part of the science of condensed matter. The surfaces of solids, however, have been more difficult to study than bulk solids for several reasons. For one, many surfaces having differing structural, chemical, and electronic properties can be cut from the same bulk crystalline material. For another, providing an atomically clean surface and keeping it clean long enough to do an experiment requires a vacuum several factors of 10 better than those usually available in the laboratory. Thus, surface studies, although under way for as long as bulk studies, have lagged well behind them until recently. Still, basic knowledge of simple surfaces and surface interactions is a necessary prerequisite to understanding in atomistic terms the structure and behavior of complicated and technologically very important surfaces.

Recent growth in surface studies has been ascribed by many to the development of convenient and commercially available vacuum pumps, some of new design, as well as analytical tools and metal parts from which an ultrahigh vacuum apparatus can be constructed. Equally, or perhaps more important, surface experiments performed in glass apparatus and the theory of solid surfaces had begun to demonstrate that a basic surface science was possible. Also, the growth of surface science has been profoundly affected and accelerated by the vital technological importance of surfaces.

More than one definition of "surface science" is possible. Viewed broadly, the term would include all work that has contributed in any way to our understanding of surfaces and surface processes. Much of this work has been empirical and not coupled to detailed atomistic understanding; nevertheless, it has very great scientific and technological value. We will discuss here, however, those investigations in surface science that are attempting to achieve understanding at the atomic level, involving the concepts of quantum mechanics where appropriate. It is this aspect of surface science that has blossomed during the past 15 years. The work is aimed at understanding and characterizing the chemical, geometrical, electronic, and motional properties of the atoms that make up the outermost few layers of a solid. One objective is to extend such information to solid–solid and solid–liquid interfaces where possible. Of course, such information about the nominally static surface is obtained by interacting with the surface in some way. But, beyond this, surface scientists want to understand at the atomic level the interactions of surfaces with molecules from the gas phase and the interactions among atoms and molecules adsorbed to the surface.

SURFACE EXPERIMENTATION

The experimental investigation of surfaces poses a number of peculiar problems. Surface probes that can distinguish the surface from the underlying and intimately connected bulk must be devised. Surface experiments must be performed in ultrahigh vacuum—that is, in a pressure of residual gases in the range of 10^{-8} Pascal (Pa), or one tenth of a millionth of a millionth of an atmosphere (atm). The surface scientist must prepare, analyze, and study surfaces in ultrahigh vacuum because an atomically clean surface, once achieved, is usually highly reactive chemically and so is covered quickly with layers of adsorbed atoms or molecules. Thus, research on surfaces takes on a quite different aspect from much of the research on bulk solids. If we imagine creating two surfaces by cleaving a solid along an appropriate plane, we realize that chemical bonds between what were bulk but now have become surface atoms are broken. These broken bonds provide the high chemical reactivity of atomically clean surfaces.

Thus, the first task of the surface scientist is to prepare the surface to be studied in a vacuum of such quality that the prepared surface will remain essentially unchanged during the course of the experiment. An atomically clean metal surface in a pressure of adsorbable gas of 4×10^{-9} Pa, or 4×10^{-14} atm, will adsorb approximately 10 percent of a monolayer in one hour. Clean semiconductor surfaces are less reactive by a factor of about 10 or more. In many experiments, the surface to be studied will involve the formation of a two-dimensional mesh of foreign atoms adsorbed to the surface. The mesh must be formed by admitting the requisite gas or vapor quickly enough so that the final surface structure does not include an appreciable amount of unwanted foreign material from the omnipresent background gases. In many cases, such a superstructure will satisfy the broken surface chemical bonds at the surface, resulting in a much less reactive surface that requires reconstruction perhaps only every few hours.

SURFACE COMPOSITION

The chemical identity of an atom is determined by the electric charge on its nucleus. The nuclear charge, in turn, bears a one-to-one relationship to nuclear mass, neglecting the complication of isotopes. These principles are the basis of three methods for determining the chemical composition of surfaces. Each method is sensitive to approximately 1 atom per 1,000 surface atoms.

One of the methods involves bombardment by high-energy (1 megaelectronvolt—MeV) helium ions, which strike and are scattered by atoms in the surface. The kinetic energy of a helium ion before and after collision, and the angles of incidence and scattering, uniquely determine the nuclear mass and thus the chemical identity of the atom struck. This method, known as ion scattering spectroscopy (ISS), is nondestructive and provides a quantitative measure of chemical composition in the outermost monolayer of surface atoms.

The other two methods determine nuclear charge, and thus chemical identity, by measuring the kinetic energies of electrons ejected from surface atoms. These electrons, when in the atom, are core electrons—those that orbit closest to the nucleus of the atom. In X-ray photoelectron spectroscopy (XPS), also known as electron spectroscopy for chemical analysis (ESCA), the electron is ejected by absorption of an X-ray photon of known energy. In Auger electron spectroscopy (AES), the electron is ejected in a process of electron rearrangement in the atom that follows the removal of a tightly bound core electron by high energy (3 kiloelectronvolts—keV) electron bombardment. Each of these methods allows one to deduce the spectrum of binding energies of atomic core electrons. Binding energy—the strength of the bond that ties the core electron to the nucleus—is the fingerprint of chemical identity. Conveniently engineered apparatus for identifying surface atoms by XPS and AES are available commercially and used widely, both in basic scientific experiments and in technological applications.

SURFACE GEOMETRY

The surface scientist is vitally interested in the relative positions of atoms in a surface, whether it is crystalline or amorphous. The surfaces of crystalline solids are ordered in repetitive arrangements of atoms in the two dimensions of the surface plane but not in the direction perpendicular to the surface. Amorphous surfaces display no long-range order in any direction, but the local structure around a surface atom is nevertheless of interest and can be determined. Detailed knowledge of surface geometrical structure is fundamental to any interpretation at the atomic level of surface electronic and chemical phenomena.

Atoms at all solid surfaces are shifted from the positions they would occupy in the bulk solid. Such surfaces are said to be reconstructed. The simplest reconstruction occurs at many clean metal surfaces where the atomic arrangement is that of the bulk solid up to and including the outermost layer except for the movement of this layer as a whole perpendicular to the surface, usually toward the bulk. Metal surfaces adsorb foreign atoms preferentially into surface sites, called hollow sites, where the number of nearest-neighbor metal atoms in the substrate is maximized. The distance between the adsorbed atom and a nearest neighbor metal atom, the bond length, tends to be near that in bulk compounds or in molecules made up of these atoms.

The geometry of the surface of a clean semiconductor, an example of a nonmetal, differs significantly from that of the bulk. This large reconstruction occurs because semiconductor lattices are open structures with the atoms farther apart than in metals, making energetically favorable rearrangements of the broken but directed atom-to-atom bonds at the surface more probable. Displacement of atoms from the positions they normally occupy in the bulk solid can be large and involve atoms several layers into the crystal.

A partial overlayer of foreign atoms adsorbed to a solid surface (atomic hydrogen on tungsten, atomic oxygen on nickel, for example) can assume different ordered arrangements called two-dimensional surface phases. The study of transitions among these phases, as determined by surface coverage and temperature, is a fascinating addition to a growing field loosely known as two-dimensional physics.

Order–order transitions between ordered phases are observed, as are order–disorder transitions involving the "melting" of the two-dimensional crystal. Some clean surfaces of metals and semiconductors also undergo phase transitions as temperature is varied with no foreign atoms present. Some phase transitions between ordered lattices of adsorbed foreign molecules also involve displacive reconstruction of the surface atoms of the metal substrate. In more weakly bound systems, one example being helium atoms adsorbed on graphite at low temperature, we observe commensurate phases, those in registry with the structure of the underlying substrate, as well as incommensurate phases and transitions between them. Studies of surface phases and transitions among them provide information on the nature of long-range ordering and melting in two dimensions and on how these are related to the characteristics of the forces between atoms within the two-dimensional layer and the forces between the adsorbed atoms and the substrate.

The surface scientist now has available a variety of methods for determining atomic positions in the surface regions of solids, both metals and nonmetals. Some are discussed briefly here.

The most widely used of these methods is electron diffraction. The two versions of the method are low-energy (50–200 electronvolts—eV) electron diffraction (LEED) and high-energy (20–50 keV) electron diffraction (HEED). In either, a beam of electrons strikes and is diffracted or reradiated by the surface. The diffracted beams produce characteristic patterns of spots or lines on a phosphor screen. From these patterns, one can determine the symmetry and size of the basic repetitive unit from which the geometrical arrangement of atoms at the surface is constructed. The method also can detect surface defects like those shown in figure 2. The variation of the intensity of low-energy diffracted beams with the energy of the incident electrons is compared with computer calculations based on specific structural models to obtain interatomic distances. Low-energy electron diffraction is the method used most frequently for structural studies of clean surfaces and studies of chemical reactions on surfaces. High-energy electron diffraction is ideal for observing the surface of a growing crystal, as in the fabrication of electronic devices by molecular beam epitaxy. Dif-

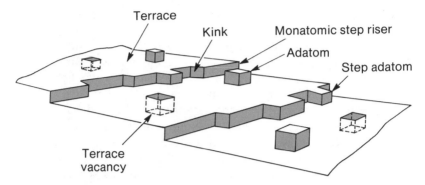

Figure 2 A schematic representation of a stepped surface with several kinds of structural irregularities. Atoms are represented by cubical spaces. The step tread, or terrace, may have atomic vacancies in its top layer or additional adsorbed atoms (adatoms) above this layer. The step riser can have kinks and attached step adatoms. [SOURCE: After G. A. Somorjai. "Catalysis and Surface Science," *Surface Science*, Vol. 89, No. 1–3 (November 1979), p. 506.]

fraction from surfaces of positrons, as opposed to electrons, also has been observed and studied.

The scattering from surfaces of atomic particles, such as helium ions (He$^+$) at high energy (1 MeV) and neutral helium atoms at thermal energies (60 MeV), also can provide information on surface geometry. Scattering of a fast ion beam aligned along a specific direction in a crystal can detect the displacement of surface atoms from the bulk positions. Using a shadowing technique, the position of an adsorbed atom above a surface can be determined by finding the direction of the incident beam that causes the emerging scattered beam, which is scattered from a substrate atom, to be blocked by the adsorbed atom. Neutral helium atoms scattered from semiconductor and adsorbate-covered metal surfaces produce diffracted beams whose angular variation with direction of the incident beam depends on the arrangement of the surface atoms of these atomically corrugated surfaces. Specific adsorption geometries have been identified in this way.

Structure in the immediate neighborhood of a surface atom can be determined by a method called extended X-ray absorption fine structure (EXAFS). This technique depends on the ejection of electrons, by X-ray absorption, from foreign surface atoms such as iodine adsorbed on a metal surface. As a result of the scattering of the ejected electrons by neighboring metal atoms back toward the "central" iodine atom, the probability of X-ray absorption varies periodically as the energy of the X-ray is varied. Interpretation of this variation determines both the number of nearest-neighbor metal atoms, which identifies the iodine adsorption site, and the length of the bond between the iodine and these metal atoms.

The diffraction of X-rays from solids is the most powerful tool for determining bulk crystalline structure. Its application to surfaces is made difficult by the weak interaction of X-rays with surface atoms and the consequent masking of the surface diffraction by diffraction from the bulk. Recently, however, it has been shown that the structures of ordered surface layers of either clean or adsorbate-covered substrates can be determined using X-rays that strike the surface at a glancing angle—less than one degree off parallelism with the surface plane. The weakness of the interaction is then an advantage, since interpretation is easier than for a strongly interacting probe, such as the electron, for which multiple scattering events must be considered.

Theoretical calculations of parameters of crystalline semiconductors have yielded information on semiconductor surface geometry. One such parameter is total crystal energy. The procedure is to calculate this energy as a function of the positions of surface atoms and to vary these positions until a minimum energy, which the crystal normally assumes, is demonstrated. Specific positions of surface atoms and interatomic bond directions, which differ appreciably from the corresponding features in the bulk, have been determined for clean surfaces of both elemental and compound semiconductors such as silicon, germanium, gallium arsenide, and indium phosphide. Another parameter is the distribution in energy of the atomic electrons in the surface region, the surface electronic structure (see below). Comparison of experimental measurements of this distribution with theoretical calculations for a variety of surface geometrical models also yields information on the positions of surface atoms.

Some methods of removing or desorbing atoms from surfaces provide structural information. Bombarding with electrons or light photons can in some instances desorb ionized atoms in preferential directions that relate to the geometry of the bonding of the atom to the surface. Understanding of the detailed nature of the electronic processes that stimulate such desorption also illuminates the character of the chemical bonding.

SURFACE ELECTRONIC STRUCTURE

The electronic structure of any atomic system denotes the distribution of its electrons both in space and in energy. Distribution in space of electrons of all energies is the spatial electron density; it is measured in electrons per unit volume. Distribution in energy of electrons in a specified spatial region (the surface region, for example) is the number of electrons per unit energy range at a given energy. This quantity is called the energy density, or simply the density, of electronic states. The energy here may be thought of as the electron's binding energy, the energy required just to remove it from the atomic system.

The most weakly bound electrons in a solid, the valence electrons of the constituent atoms, are those that provide the chemical bonding that holds the atoms of the solid together. The electronic structure involving these electrons affects profoundly the chemistry that can occur at the solid surface and the electronic characteristics of surfaces and interfaces in electronic devices. We must understand surface electronic structure if we are to design surfaces having specific electronic characteristics. Catalytic activity, for example, has been correlated empirically with the electronic characteristics of the catalyst. Control of surface electronic states that act as electron traps is essential to the successful design of electronic devices.

The valence electron states of a solid are arranged in bands on the scale of binding energy; the bands are separated by forbidden energy gaps in which no electronic states exist. The electronic structure of a bulk solid is significantly altered in the surface region, even for clean surfaces harboring no foreign atoms. For example, the surface states mentioned above confine electrons to the surface region at energies in the forbidden gaps of the bulk solid. Foreign atoms attached to surfaces also modify the surface electronic structure significantly.

Surface experimentalists determine surface electronic structure by various electron spectroscopies, in which electrons are ejected from the solid by an external agent. The kinetic energy distribution of the ejected electrons, and in some cases also their angular distribution, are measured. The most important and commonly used agent is the light photon, which is employed in a method known as photoelectron spectroscopy. Synchrotron radiation, because of its variable energy, polarization properties, and high intensity, is the most versatile source of photons. Other electron spectroscopies involving the use of high electric fields, incident ions, or metastably excited atoms also have been developed. Energy loss of scattered electrons and absorption of light yield information on the electronic characteristics of surfaces. Theorists have developed powerful methods of calculating both the spatial density of electronic charges and the density of electronic energy states both at surfaces and at interfaces.

Core electrons of atoms in a solid have almost the same binding energies as they do in a free atom. However, the small shifts that do occur on incorporation into the solid are significant because they indicate the electrostatic character of the chemical bonds to the atom. The same applies to surface atoms, where altered core-level shifts indicate conditions peculiar to the surface.

SURFACE ATOMIC MOTIONS

The motions of surface atoms of interest to the surface scientist are of two types, diffusive motion and vibrational motion. An example of diffusive motion is the movement of atoms across a surface. Such motion has been studied extensively by both field emission and field ion microscopies. The latter can image the surface with atomic resolution. Diffusive motion also may be taken to include passage through the outermost layer of atoms to positions in, for example, the second or deeper layers. Evidence for such motion is obtained using electron spectroscopies that determine chemical composition and the structure of electron energy levels. Of great importance in surface science is the diffusion of impurity atoms from the bulk into the surface region of the solid and their segregation there.

We may distinguish two types of surface atomic vibrations. For a clean surface, there are modes of periodic motion about an equilibrium position. The frequencies of these vibrations can be measured by the absorption of infrared light and by the transfer of energy and momentum to surface vibrations by scattered electron, atomic, or molecular beams at the surface. Infrared absorption and high resolution electron energy loss also yield information on the frequencies of the several possible modes of vibration of a foreign atom or molecule adsorbed to the surface. Determination of the frequencies of several of the possible modes of vibration is sufficient in many cases to determine the bonding structure of the atom or molecule at its adsorption site.

SURFACE SCIENCE AND TECHNOLOGY

Basic knowledge of surfaces and interfaces is critically important in several technologies based on surface or interface phenomena. Here, three examples of the impact of surface science on technological problems will be discussed.

The first concerns a technology of crystal growth that developed directly out of studies in surface science and that could not exist before surface-science techniques were perfected. This technology, molecular beam epitaxy (MBE), has made possible a number of startling innovations in crystal growth, interface formation, and surface treatment leading to several new electronic and electrooptical devices.

The second involves the impact of surface science on a well-established and vitally important technology, heterogeneous catalysis of chemical reactions. This technology is highly developed on the basis of more macroscopic methods of study and procedure, in many cases involving a large element of empiricism. Several important steps are being taken to bridge the gap between basic surface science and catalyst design and operation.

The third describes the widespread application of surface analytical techniques to a variety of technological problems. An example of such a problem is the deleterious segregation of impurity atoms at surfaces, interfaces, and grain boundaries of technologically important materials.

Molecular Beam Epitaxy

Epitaxy is crystal growth in which a crystalline substrate determines the crystallinity and orientation of the layers of atoms grown upon it. The materials needed for crystal growth are most commonly deposited from vapors or liquids in processes called vapor-phase and liquid-phase epitaxy. In molecular beam epitaxy, however, all of the atoms needed to grow multiatomic crystals are supplied via molecular beams. The beams are produced by effusion from ovens in a cold-wall, all-metal vacuum system, pumped to ultrahigh vacuum and equipped with basic surface characterization tools. Molecular beam epitaxy has been demonstrated to be a versatile technique, capable of extreme dimensional control that makes possible the growth of layered structures with precise electrical and optical properties.

The essential components of an MBE apparatus are shown schematically in figure 3. In an ultrahigh vacuum apparatus, heated effusion ovens, as many as six, produce beams of the constituent atoms incident on the heated substrate. Each beam is controlled by a shutter. Achievement of high crystalline and semiconducting quality requires the maintenance of a clean vacuum and a substrate sufficiently hot to make surface atoms mobile enough to migrate to the proper crystal sites. Also shown in figure 3 are the obliquely incident electron beam and fluorescent screen used in the continuous observation, by high-energy electron diffraction (HEED), of surface crystalline structure as growth proceeds. This capability is extremely important in controlling the conditions of crystal growth and is unique to MBE.

As an example of the use of MBE, consider the growth on a gallium arsenide (GaAs) substrate of layers of either GaAs itself or of the mixed crystal aluminum gallium arsenide, abbreviated (Al,Ga)As (figure 4). The growth of gallium arsenide layers on a gallium arsenide substrate requires the incorporation of equal numbers of gallium and arsenic atoms in each layer. The means of doing this grew out of the study of the interaction of gallium atoms and arsenic molecules with the atomically clean gallium arsenide surface. It was shown that the surface lifetime of gallium atoms on gallium arsenide is greater than 10 seconds at a surface temperature of about 500° Celsius, while arsenic molecules quickly desorb unless gallium is present. When both gallium and arsenic beams impinge on the hot substrate with the atomic arrival

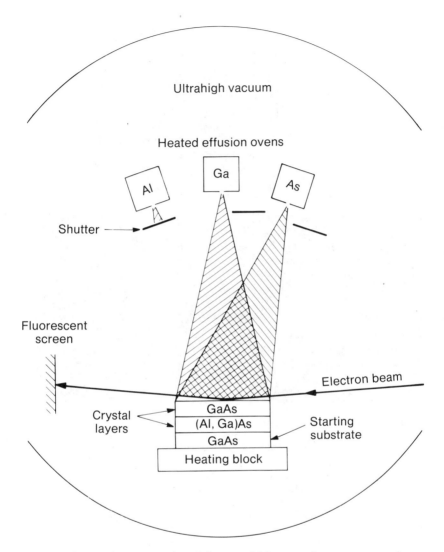

Figure 3   Schematic representation of the essential features of an apparatus used to grow epitaxial layers of gallium arsenide (GaAs) and aluminum gallium arsenide [(Al,Ga)As] by molecular beam epitaxy and to observe the growth by means of the diffraction of high-energy electrons striking the surface obliquely. [SOURCE: After A. Y. Cho and J. R. Arthur. "Molecular Beam Epitaxy," in *Progress in Solid State Chemistry*. Vol. 10, Pt. 3. New York: Pergamon Press Inc., 1976, pp. 157–191.]

rate of arsenic greater than that of gallium, one arsenic atom will remain on the surface for each gallium atom provided. In this way, stoichiometric gallium arsenide can be grown.

The best mechanically or chemically polished GaAs substrates, when observed by electron diffraction (HEED), are seen to be rough.

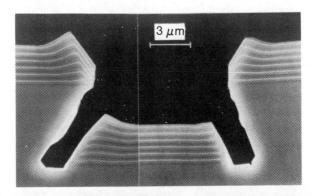

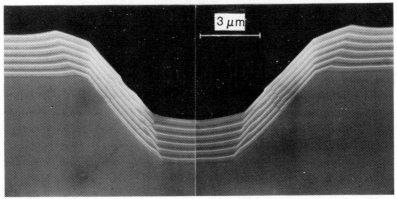

Figure 4 These pictures taken with a scanning electron microscope show the precision of molecular beam epitaxy, an emerging method for fabricating electronic and electrooptical devices. The banded areas are alternating layers of gallium arsenide and gallium aluminum arsenide. The thickness of each layer is 0.22 μm—slightly more than two-tenths of a millionth of a meter. [SOURCE: W. T. Tsang and A. Y. Cho. "Growth of Gallium Arsenide–Gallium Aluminum Arsenide ($Ga_{1-x}Al_xAs$) Over Preferentially Etched Channels by Molecular Beam Epitaxy: A Technique for Two-Dimensional Thin-Film Definition," *Applied Physics Letters*, Vol. 30, No. 6 (March 15, 1977), pp. 293–296.]

HEED can detect the roughness because the obliquely incident electron beam actually passes through the hillocks on the surface and, in the process, undergoes three-dimensional diffraction. Three-dimensional diffraction produces a pattern of isolated spots on the fluorescent screen, whereas two-dimensional diffraction from an ideally smooth surface produces a pattern of easily distinguishable vertical lines. An extremely important aspect of MBE growth is that it smooths the surface as it proceeds, with the newly condensed constituents filling the valleys of the rough surface.

A second important characteristic of MBE is that auxiliary effusion ovens can be used to incorporate impurity elements into otherwise stoichiometric layers. Thus, one can control the electronic conduction mechanism of any given layer. Incorporation of impurity elements (dopants) having a greater or smaller number of valence electrons than are needed in the chemical bonding produces "n" type conduction layers, in which conduction occurs via electrons, or "p" type layers, in which it occurs via holes (electron vacancies). What is more, the concentration of the impurity atoms can be controlled precisely by the intensity of the molecular beam and can be varied through a given layer according to a specific profile. When the relative intensities of the constituent beams striking the substrate are changed suddenly, an abrupt junction is formed. In such a junction, the electronic and optical characteristics of the material change in a very short distance.

The control of crystal growth provided by molecular beam epitaxy is such that a layer of gallium arsenide that is 100 atoms thick will vary in thickness by only one or two atoms over the entire layer. MBE is the first crystal fabrication technique to achieve this level of control. Gallium, arsenic, and dopant atoms also may be deposited on a heated substrate in a method known as metal organic chemical vapor deposition (MOCVD). This deposition occurs as part of a chemical reaction among metal-organic molecules at the hot substrate. The method shows great promise for growing thin crystalline layers in which composition and thickness are controlled precisely.

Single-crystal, multilayered structures with component layers differing in composition but not in geometrical lattice structure form the basis for semiconductor devices in which both light and current carriers can be manipulated. The double heterostructure laser is an example of a three-layer device. In it, electrons from one layer combine with holes in the adjacent layer to produce light, which is then channeled in the adjacent layer.

Multilayer structures of alternating, sharply defined layers of semiconductor material of differing properties form what are known as superlattices. In one such structure, alternate layers of aluminum gallium arsenide containing impurity atoms that donate conduction electrons, are interleaved between layers of pure gallium arsenide. Electrons donated from the doped (Al,Ga)As layer "spill over" into the pure GaAs layer, where their velocity at low temperature can be 100 times greater than in material uniformly doped to a comparable concentration level, where the electrons are slowed by scattering off the atoms that donated them. This new, high-mobility material, with donor atoms and donated electrons separated, is useful in the design of high-speed electronic devices. Gallium antimonide (GaSb) and indium arsenide (InAs) also have electronic structures sufficiently different that alternate, sharply defined layers of the two will produce a high-mobility compositional superlattice, but without the use of specific dopant atoms. Electrons in the GaSb layers spill over into the InAs layers, while holes migrate from InAs to GaSb.

The motion of electrons in the extremely thin layers of multilayer materials like those discussed above is of great interest to the fundamental scientist. Here, he has the opportunity to observe and study the behavior of a two-dimensional electron "gas" whose properties differ startlingly from those of the "gas" of electrons found in three-dimensional solids. Thus, technology that has produced new materials with the aid of basic science has in turn enriched basic science.

Heterogeneous Catalysis

A catalyst is a substance that accelerates a chemical reaction without itself being consumed in the process. In a common type of heterogeneous catalysis, the catalyst is a specially prepared solid surface that enhances the chemical reaction of gaseous constituents. The general character of the sequence of steps involved is known. Gaseous molecules are adsorbed on the catalyst, where they can be broken into fragments. The adsorbed molecule or its dissociated fragments diffuse over the surface of the catalyst and react chemically with it or with other species adsorbed from the gas phase. These processes produce surface intermediate species and, finally, a product species that is desorbed from the catalyst into the gas phase. The design of catalysts by chemists has involved ingenious approaches based on both physical and chemical knowledge and intuition. It has included careful measurements of reaction rates and specificity, as well as the characterization of the catalyst so designed, using a variety of physical and chemical methods. Because of the complexity of the problem, these semiempirical methods will continue to be important design modes.

Basic surface science has contributed much in recent years to our understanding of the various steps or component processes involved in catalytic reactions. Adsorption and desorption of molecules from surfaces have been characterized at the atomic level. Structures formed by adsorbed materials have been determined. Molecular rearrangements occurring on surfaces also have been investigated. All of this work that has enhanced our atomistic understanding of catalytic processes has been done in the low-pressure range (around 10^{-8} Pa, or 10^{-13} atm), where the new analytical tools operate. Commercial catalysts, on the other hand, operate at pressures in the millions of Pascals (tens of atmospheres). This pressure gap must be bridged if our basic understanding of catalysis is to proceed.

Experimental apparatus has been developed in which it is possible, at low pressure, to characterize the surface to be used as a catalyst

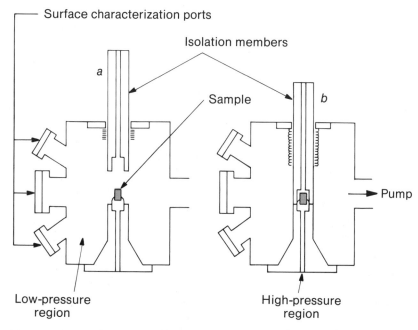

Surface characterization ports

Isolation members

a

Sample *b*

Pump

Low-pressure region

High-pressure region

Figure 5 Schematic representation of an apparatus in which a surface can be characterized by several low-pressure techniques (configuration at left) and also operated as a catalyst in a high-pressure chemical reaction chamber (configuration at right). [SOURCE: After G. A. Somorjai. "Catalysis and Surface Science," *Surface Science*, Vol. 89, No. 1–3 (November 1979), p. 497.]

and, at high pressure, to perform a catalytic reaction. The essential features of one such apparatus are shown schematically in figure 5. At the left, with the isolation member at position *a*, the model catalyst, a single crystal of small surface area (1 square centimeter), can be characterized by surface-science techniques. Samples of differing material and of various surface crystallographic orientation may be employed. Stepped surfaces with either linear or kinked risers between the terraces or treads, as illustrated in figure 2, have been studied.

The high-pressure catalytic reaction is performed in the small enclosure about the sample surface formed by lowering the isolation member to position *b*, as seen at the right in figure 5. Reactant gases at pressures as high as 10^7 Pa (100 atm) may be circulated over the sample and the changes in gas composition determined by standard techniques as the catalytic reaction proceeds. Meanwhile, the low-pressure region is kept at low pressure. The catalytic reaction may be stopped at any time, the gases pumped out, and the isolation cell opened, permitting the catalyst to be analyzed again at low pressure. Structural modification and adsorption on the catalyst's surface can then be evaluated.

This apparatus has been used, for example, to study hydrocarbon reforming by which high-octane gasoline is made. It also has been used to study modification by hydrogen addition to produce new hydrocarbons. It has been shown that reaction rates on platinum surfaces are sensitive to specific surface structural characteristics. The rates are faster on some crystal faces than on others and are affected by such surface structural irregularities as those illustrated in figure 2. There is evidence that surface steps, for example, can affect the ability of a surface to dissociate certain chemical bonds as molecules adsorb. Chemical changes on surfaces, such as the formation of a carbonaceous layer on platinum, are important in determining catalytic reaction rates. Similar sensitivity to surface geometrical structure has been demonstrated in the iron-catalyzed synthesis of ammonia from hydrogen and nitrogen.

Polymetallic cluster catalysts have an important place in petroleum refining. Recent advances in understanding the structure of such highly dispersed systems have come through the use of high-resolution electron microscopy and extended X-ray absorption fine structure in conjunction with the more traditional use of selective chemisorption and model catalytic reactions as probes. In some cases, one component is found to segregate to the surface of the cluster and thus to dominate its chemical behavior. Studies using model catalysts constructed by adsorption of one component on the crystalline surface of another corroborate this conclusion.

Thus, basic surface studies are providing extensive new knowledge of the various individual atomic and molecular processes that together constitute the phenomenon of heterogeneous catalysis. Gaps in our knowledge persist, however, particularly because we cannot apply atomic surface analysis techniques during the high-pressure catalytic process. Still, it is not unreasonable to expect that the knowledge surface science is providing at the atomic level about surface chemical reactions will lead to improved catalyst design, which hitherto depended in many instances on semiempirical methods.

Surface Analysis

One of the most important influences of basic surface science on technology has come from the development of surface chemical analytical tools with spatial resolution of a few atomic monolayers. These are the techniques for chemical identification discussed above. Primary among them is Auger electron spectroscopy. AES is sensitive to the outermost few atomic layers, can detect elements of both high and low atomic number (except hydrogen) to the limit of 0.1–0.5 atomic percent, and can be arranged to have good spatial resolution at the mi-

Branching aggregates of silver sulfide formed by exposing silver to air containing sulfur. Photograph taken with a scanning electron microscope. [Bell Laboratories.]

crometer level. When combined with the slow removal of material by ion bombardment (sputtering), AES provides a means of profiling chemical constitution to near atomic resolution through bulk material, including solid–solid interfaces of junctions in semiconductor devices.

Secondary ion mass spectrometry (SIMS), in which ions ejected from the surface by primary ion bombardment are mass-analyzed in a mass spectrometer, also has technological uses in detecting trace elements and in analyzing the surfaces of insulators. X-ray photoelectron spectroscopy (XPS or ESCA) has the requisite surface sensitivity and provides both elemental chemical analysis and information on the chemical state of the atom analyzed. Ion scattering spectroscopy has spatial sensitivity to the outermost layer of atoms only and can be used for elemental analysis of both conductors and insulators. SIMS and ISS have not found as wide usage as AES and XPS, but they have their own unique and useful characteristics.

With particular emphasis on Auger electron spectroscopy, it may be said that the capabilities of elemental analysis with a depth resolution of a few monolayers of atoms have had a profound effect in several fields. The three most important of these are solid-state electronic devices, metallurgy, and catalysis.

Depth profiling provides the means of evaluating the success or failure of several semiconductor processing procedures. Among them are the generation of specific impurity chemical profiling, interface or junction formation, and surface passivation. Failure of devices is traceable in many instances to the incorporation of unwanted impurities at specific and determinable positions in the device. In metallurgy, surface analytical techniques have affected profoundly the elucidation of diffusion and segregation at internal interfaces, which lead to embrittlement and structural failure. Fractured surfaces can be analyzed and intergranular corrosion phenomena studied. Similarly, spatial chemical composition and modification of newly designed catalysts can be evaluated, and possible poisoning (inactivation) of catalysts can be evaluated after their use in catalytic reactions.

Outlook

Surface science has made great strides during the past decade or two in its ability to provide detailed understanding at the atomic level. But much remains to be done, both to continue the development of the science itself and to provide a strong base for the important technologies in which surface phenomena figure prominently. Successful continued development of these technologies is vital to the economic growth, competitive abilities, and security of the United States.

One of the most important tasks facing the surface scientist is to build detailed understanding of chemical reactions at surfaces. The preparation of chemical reactants in specific vibrational and rotational states and the resolution in time of surface processes are new means to this end. Considerable refinement of our techniques for determining the geometrical, electronic, and vibrational structures of both the surface and the species adsorbed on it is in the offing. Improvement in catalytic and energy efficiency can be expected, for example, in such diverse catalytic processes as the production of fertilizers, polymers, refined petroleum, and synthetic fuels.

Another important facet of surface science concerns the role of the geometrical and electronic properties of surfaces in the design, fabrication, and operation of high-speed integrated electronic circuits. The technology will involve further development of multilayer structures having high electron mobility, interfaces or junctions of improved electrical properties, and means of rendering specific surfaces of the new materials passive. Central to an emerging high-speed integrated circuit technology is the fabrication, by molecular beam epitaxy, of transistors involving well-defined layers of differing atomic composition that are thin enough to permit appreciable spillover of electrons between layers. This new technology will provide logic circuits for higher speed computers, for example. Similarly, optical communication systems will be improved by the extension of MBE techniques to produce semiconducting structures required for lasers and optical detectors of previously unobtainable but desirable characteristics.

Our understanding and control of many metallurgical problems of great import to society will be advanced by the use of surface analytical tools. Chemical corrosion, as an example, is a very costly phenomenon occurring on bridges and in nuclear reactors. Detailed knowledge of the process is essential to the design of corrosion-resistant materials or corrosion-preventing coatings. Expensive, empirical, trial-and-error solutions must give way to ideas based on understanding of the relevant phenomena at the atomic level.

The science of surfaces and interfaces is a multidisciplinary endeavor fundamental to many important technologies in addition to those touched on in this brief look at the future. It tends to break down walls between physicists, chemists, metallurgists, and technologists. Often, it is difficult to detect what is science and what is technology and which of these in any particular instance is more important to the enrichment of the other. In any event, it is essential that the United States maintain a position in the forefront of scientific and technological development in this broad, vital, and expanding field.

BIBLIOGRAPHY

W. F. Brinkman et al. "Melting of Two-Dimensional Solids," *Science*, Vol. 217, No. 4561 (August 20, 1982), pp. 693–700.

A. Y. Cho. "Recent Developments in Molecular Beam Epitaxy (MBE)," *Journal of Vacuum Science and Technology*, Vol. 16, No. 2 (March–April 1979), pp. 275–284.

R. Dingle et al. "Building Semiconductors From the Atom Up," *Bell Laboratories Record* (September 1980), pp. 274–281.

D. E. Eastman and F. J. Himpsel. "Ultraviolet Radiation—An Incisive and Versatile Tool," *Physics Today*, Vol. 34, No. 5 (May 1981), pp. 64–71.

G. Ehrlich. "Wandering Surface Atoms and the Field Ion Microscope," *Physics Today*, Vol. 34, No. 6 (June 1981), pp. 44–53.

P. Eisenberger and L. C. Feldman. "New Approaches to Surface Structure Determinations," *Science*, Vol. 214, No. 4518 (October 16, 1981), pp. 300–305.

R. Gomer. "Surface Diffusion," *Scientific American*, Vol. 247, No. 2 (August 1982), pp. 98–109.

H. D. Hagstrum. "Surface Physics," in *Physics Vade Mecum*. Edited by H. L. Anderson. New York: American Institute of Physics, 1981, Section 21.00, pp. 300–313.

I. Lindau and W. E. Spicer. "Photoemission as a Tool to Study Solids and Surfaces," in *Synchrotron Radiation Research*. Edited by H. Winick and S. Doniach. New York: Plenum Publishing Corp., 1980, pp. 159–221.

G. E. McGuire and P. H. Holloway. "Applications of Auger Spectroscopy in Materials Analysis," in *Electron Spectroscopy: Theory, Techniques, and Applications.* Volume 4. New York: Academic Press, Inc., 1981, pp. 1–84.

M. B. Panish. "Molecular Beam Epitaxy," *Science*, Vol. 208, No. 4446 (May 23, 1980), pp. 916–922.

M. Prutton. *Surface Physics*. Oxford: Clarendon Press, 1975.

J. R. Schrieffer and P. Soven, "Theory of the Electronic Structure"; D. E. Eastman and M. I. Nathan, "Photoelectron Spectroscopy"; R. L. Park, "Inner-Shell Spectroscopy"; E. W. Plummer et al., "Vacuum-Tunneling Spectroscopy"; *Physics Today—A Special Issue on Surface Physics*, Vol. 28, No. 4 (April 1975), pp. 23–71.

J. H. Sinfelt. "Structure of Metal Catalysts," *Reviews of Modern Physics*, Vol. 51, No. 3 (July 1979), pp. 569–589.

N. V. Smith and D. P. Woodruff. "Surface Spectroscopies with Synchroton Radiation," *Science*, Vol. 216, No. 4544 (April 23, 1982), pp. 367–372.

G. A. Somorjai. *Chemistry in Two Dimensions: Surfaces*. Ithaca: Cornell University Press, 1981.

6

Turbulence in Fluids

Fluids in motion, whether gases or liquids, play a vital role in human affairs: in the performance of aerodynamic structures and ships; in the oil flowing through pipelines from the North Slope of Alaska; in the pollution of our air and rivers; in the restraints on our activities and occasionally the threat to our lives imposed by the vagaries of atmosphere and ocean; even in the drifting continents as they fuel and trigger a Mt. St. Helen's eruption. Fluids in motion also are characteristically turbulent. Thus, an understanding of turbulent flow is essential in many endeavors, both scientific and technological.

What we now know of turbulent flow is almost entirely empirical. In more than 100 years of scientific study, negligibly few quantitative predictions of turbulence have been deduced from theory. But recent discoveries in experiment and theory have created promising new directions for research on the nature of the turbulent process in fluids. Progress will not come rapidly, but the stakes are high. More basic understanding of turbulence should lead to marked gains in practical activities that include the design of aircraft, compressors, turbines, chemical reactors and mixers, and many other products; prediction of flow factors for offshore and onshore structures; and improved prediction of weather and ocean currents.

Fluid dynamicists define turbulence as continuously disordered flow. Inside a turbulent air mass or river, for example, the speed and

---

◀ Both the pattern and the angle of the waves on either side of the ship form as predicted. [J. N. Newman, MIT, in *An Album of Fluid Motion*, by Milton Van Dyke.]

direction of movement of the fluid at any point are changing continuously and erratically; the bulk of the fluid, meanwhile, moves on in its appointed direction. The development of turbulence in a smoothly flowing fluid is a sequential process: one or several initial instabilities trigger the onset of disordered flow, which continues to intensify until it reaches a point of high disorder as fully evolved turbulence.

A bumpy ride in an airplane and a boat that will not hold its course are manifestations of turbulence. In more technical terms, turbulence dramatically enhances the transport of particles, heat, and momentum in fluids. Such enhancement may be desirable, as in the mixing of chemicals or the transfer of heat in industrial processes, or it may be undesirable, as with the drag on fluids flowing in pipes or turbines. In any event, the ability to control or predict the effects of turbulence is an eminently practical goal.

THE ORIGINS OF TURBULENCE

Observations of turbulence have suggested two distinct problems to the fluid dynamicist. One is to understand the instabilities in a smoothly flowing fluid that lead to the onset of turbulence. The other is to understand the behavior of the fluid when turbulence is fully developed. Instabilities responsible for the transition to disordered motion can arise in shearing flow, which may be visualized as layers of fluid sliding across each other. Instabilities leading to turbulence also can be produced by heating a fluid from below. The two types of instabilities are different, and the resulting flows have come to be labeled shear turbulence and thermal turbulence. The former is illustrated by the invisible change in the speed of the wind with a change of altitude and the latter by the visibly changing structure of a summer cloud.

Our relatively poor understanding of motions in turbulent fluids, even in the simplest laboratory setups, is not due to the lack of a sound mathematical foundation. The basic fluid equations* represent local, or small-scale, averages of the underlying molecular motions and describe fluid motion quantitatively at any given point. The problem is that fluid dynamics is nonlinear—that is, changes in velocity in a fluid depend on the square of the velocity itself (see boxed term in footnote), as well as

---

*The Navier–Stokes partial differential equation,

$$\frac{\partial \mathbf{v}}{\partial t} + \boxed{\mathbf{v} \cdot \nabla \mathbf{v}} = -\frac{1}{\rho} \nabla P + \nu \nabla^2 \mathbf{v},$$

where $\mathbf{v}$ is the velocity, P the pressure, ρ the density, and ν the kinematic viscosity, completely describes the turbulent flow in an incompressible, isothermal fluid. The boxed nonlinear term is the source of instability and the demon that makes flow problems so intractable.

on pressure, viscosity, or internal friction, and on conditions at the boundary of the fluid, such as at the wall of a pipe. This nonlinearity is the cause of the complexity in space and time of even the simplest turbulent motion.

A principal consequence of the nonlinearity is to destabilize steady solutions of the basic equations. The pattern of growing, then decaying, ever changing, and differing velocities throughout a fluid in a state of developed turbulence defies detailed mathematical description. By direct numerical computation from the basic equations, one can, in principle, integrate the features of many small-scale flows in a turbulent fluid to determine the large-scale flow observed experimentally. The computations are so complex, however, that the largest computers today can barely address even the initial instability in shearing flow. Even at the swift rate of development of computers, they will not in our lifetime be able to resolve fully developed turbulence in pipes, a relatively simple case.

Statistical Theories

What computers can do well, however, is assist in testing the usefulness of statistical theories of flow. Most theories of turbulence rest on hypotheses advanced to approximate or replace the nonlinear term in the basic equations. The goal is to relate the features of large-scale flow to the averaged properties of the underlying small-scale flow. One seeks to substitute a simple, statistical model for a complicated process in the hope of discovering a theory or model that is generalizable, quantitative, and, finally, deducible from the basic equations as an approximation with a range of validity that can be established both theoretically and experimentally. Such a model then can be used to predict the effects of turbulence, within its range of validity, for a variety of engineering or other purposes.

These statistical theories of turbulence address aspects of fluid motion that may be insensitive to the details of what is actually happening in the fluid. That is, the statistical result is the average of the effects of the detailed mechanisms of turbulence. This search for statistical, as opposed to mechanistic, features of turbulent motion has been very important in setting the traditional directions of research in the field.

In the first simple models, it was proposed that the viscosity appearing in the fluid equations be replaced by an "empirically determined viscosity" presumed to incorporate the most important effects of the nonlinear term. This model and extensions of it have found considerable use in explaining some large-scale flows qualitatively. For each new application, however, such models must be recalibrated by new experiments. They can be used with confidence, for example, to interpolate features of turbulent flow in turbines of the same type but different sizes, but not to extrapolate such features in the design of new types of turbines.

A second class of statistical models emerged from the presumption that turbulent flow instrinsically was totally disordered. Here, one explored the first effects of the basic equations in imposing order on this disordered state. This approach is characteristic of the closure hypotheses discussed later in this chapter. These models, however, like those mentioned in the previous paragraph, lack sufficient formal links to the basic equations to make it possible to predict their range of validity.

Focus on Mechanisms

The distinguishing feature of the recent research on turbulence is that it focuses much more on the mechanisms of turbulent flow than do the two classes of statistical theories described above. The emphasis is on relatively simple mechanisms accessible to theoretical study. Hence, the new studies may seem to be less concerned with the immensely difficult central problem, fully developed turbulence, than with the more tractable problems around the edges. Yet, one of the new hopes is that a deeper understanding of fluid mechanism will permit the construction of much improved statistical theories.

Here, in the context of previous and continuing work, we will describe three of the discoveries that are causing the current excitement in research on turbulent flow. They lie in three different areas: the transition to turbulence in shear flow; the transition to turbulence caused by thermal instability; and the disordered behavior in simple mechanical and electrical systems that has provided clues linking transition and full turbulence. The results outlined are those that appear to concern elementary and generalizable aspects of turbulence and so may significantly affect research and development in the immediate future.

SHEAR-FLOW INSTABILITY

Traditionally, the study of turbulence has focused on the disorder observed in shearing flow. Such turbulence is responsible, for example, for increased drag on aircraft, the swift mixing of chemicals or gases, and much of the drag on fluids flowing in pipes; hence, it is an important practical problem. Consequently, a large body of experimental data has been developed and incorporated into a variety of models that describe aspects of the process for engineering purposes. Until recently, however, no theoretical developments had emerged that permitted the fundamental features of this turbulence to be determined quantitatively or suggested possible means of controlling it. The engineering models, moreover, did not lend themselves to generalization for such purposes. Thus, the search continued for statistical models that reflect more accurately the constraints imposed by the mechanism of turbulence.

Transition Spots

What are these constraints? The average properties of a fluid in shear flow near any boundary are largely the consequence of a very few violent bursts of disordered flow. Fluid dynamicists have little doubt that these bursts are evolved forms of the initial shear-flow transition spot of figure 1. The spot occurred in the interior of a closed channel 0.6 centimeter deep, 1 meter wide, and 5 meters long and filled with flowing water. It was photographed through the plastic wall of the channel, and the mean flow is from the top of the picture toward the bottom. The coherent structure of the spot was made visible to the camera by light reflected from tiny mica flakes suspended in the water and oriented by the passage of the spot. The observed wavelengths— the apparent spacing of the waves—are comparable to the depth of the channel.

Such transition spots now are believed to be prototypes of the flows responsible for most of the transfer of momentum and mixing of heat and particles throughout the fluid. Spots have been observed for several decades. The structural details of these fast and infrequent events, however, were not explored fully during most of that period because they are initiated on such a small scale and because it was clear to only a few scientists that they played a central role in the transport of heat, momentum, and particles. Only in the past few years has it become possible to begin to obtain quantitative agreement between experimental observations of transition spots and numerical calculations from the basic equations. Calculations on advanced computers, for example, now can predict certain features of the structure of the transition spot shown in figure 1. In consequence, considerable refinement in both theory and experiment can be anticipated.

These strange, wavelike transition spots appear abruptly in fluids because of disturbances in the flow. The meaning of "abruptly" depends on the characteristics of the system—it can be milliseconds, for example, as in the experiment of figure 1, or several seconds in a slow river. The spots are capable of growing very rapidly while extracting kinetic energy from the original fluid motion. In fully developed turbulence, one can picture many evolved spots interacting and decaying, with new, violent disturbances appearing infrequently to sustain the disorder. The study of such spots as isolated, nonlinear phenomena,

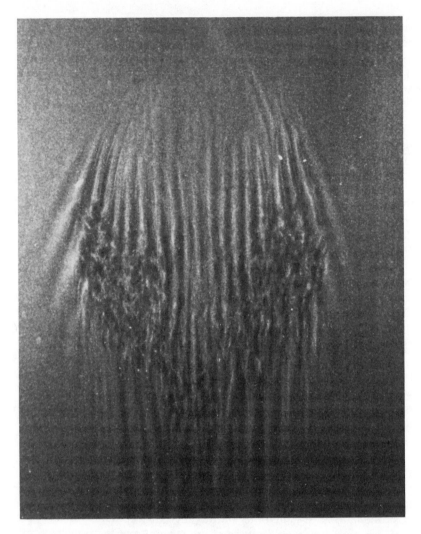

Figure 1   Shear-flow transition "spot." [SOURCE: D. Carlson et al. "A Flow Visual-ization Study of the Transition in Plane Poiseuille Flow," *Journal of Fluid Mechan-ics*, Vol. 121 (August 1982), pp. 487–505.]

therefore, seems a promising route to deeper understanding of a central feature of the turbulent process.

Even to the untrained eye, the transition spot of figure 1 suggests an analogy with the bow waves of a ship at sea. But the spot's wave system is self-propelled. The role of the ship appears to be played by the disordered flow that follows the small disturbances on the second and third bow waves. The challenge to fluid dynamicists, then, is to understand the properties of waves in shearing flow, the stability of such waves, the role of secondary instabilities, and the breakdown of waves to produce disordered flow. These problems provide the framework for a brief historical survey of work leading directly to a significant part of current research.

Wavelike Motions

The first work on wavelike motions in steady, or nonturbulent, shearing flow was initiated before 1900. Not until the 1950's, however, did a complete mathematical theory emerge for the instability of a wave in a pipe or channel exhibiting initially small departures from the average velocity. This theory correctly predicts the behavior of long, crested waves in a given steady shearing flow. The theory also predicted, however, that, under the conditions of flow of figure 1, all waves of small velocity must decay and vanish. This prediction meant that, to handle the transition to turbulence in such flow, the theory had to take account of waves of large velocity. That problem is much more difficult, and progress with it has come only during the past few years with the advent of new ideas and the help of computers.

Models of Turbulence

During the long, frustrating period of development of the theory of waves in shearing flow, many scientists and engineers evolved statistical models based on a variety of hypotheses of what actually occurs in fully developed turbulence. The first of these models advanced the view that the ordered transfer of mass, momentum, and heat in a fluid would occur as a small departure from a basically disordered flow. The simplest realization of this idea is an eddy viscosity coefficient. It is based on the concept that many small-scale eddying motions in a fluid would disrupt flow sufficiently to result in the equivalent of a large-scale viscosity.

The eddy viscosity coefficient is a statistical approximation chosen to represent the observed behavior of eddy fluctuations on the average flow. Starting with the eddy viscosity concept, extensive modeling permitted by numerical solutions of linear differential equations has led to various useful results. In general, the approach provides a framework for codifying experimental data; the calculations can adequately match flows observed experimentally in many practical engineering conditions. Hence, engineering design handbooks contain a variety of numerical models of this kind suitable for special purposes—for example, designing aircraft or fluid-handling systems (such as pumps). Such models have useful predictive value even in physical circumstances somewhat different from those of the experiments used to corroborate them. The results of models based on the eddy viscosity concept, however, have not provided a quantitative basis for a deeper understanding of turbulence.

More fundamental attempts to comprehend statistical aspects of turbulent shearing flow have been based on the mathematical idea of closure. The word describes a method of solving the sequence of equations relating average properties of the flow to other average properties. Each equation in the sequence requires knowledge of further, unknown averages, and huge numbers and types of solutions are possible. Closure entails a hypothesis that permits the sequence of equations to be terminated at a manageable number. Closure is a useful technique if the hypothesis is a good approximation, but in no instance has a mathematical way to test its validity been built into a closure procedure. The tests used for most closure schemes are the experiments. Since the schemes often contain a number of adjustable constants, it is not easy to rule out closures that are not compatible with physical laws.

A type of closure called second-order closure has become popular in recent years and has found several promising applications. One example is the modeling of the flow of pollutants in the most turbulent layer of the atmosphere—from the ground to 1 kilometer above the earth's surface. However, quantitative application of second-order closures to controlled shear flows in the laboratory has not yet succeeded.

Perhaps the most profound closure—one that is quantitative in principle—is called the direct-interaction approximation. The approximation is not quite correct in its description of very small-scale, small-amplitude flows. Still, it offers hope of a satisfactory description of the low-order (mathematically less complex) statistics of large-scale turbulent flow. The direct-interaction approximation has been rediscoverd frequently, recently by theoreticians seeking to describe turbulence in fusion plasmas. The approximation is extremely complicated even in recently improved and simplified versions. Already, however, it has been used with high-speed computers to provide sound estimates of the decay of turbulence in fluids. Faster computers and further simplification will permit additional tests of its quantitative accuracy and limits of applicability.

An extreme statistical idealization of turbulent flow is called isotropic homogeneous turbulence, in which the average properties of the flow are the same along any axis. It is difficult to produce such flow in the laboratory, even approximately. Yet, many scientists have found in its symmetries the hope of discoveries that would be universal in their simplicity and applicability. The implications of this idealization have been explored at some length. The most significant result, however, follows from the physical assumption that turbulent energy passes through intermediate scales, or dimensions, of motion in cascading from large-scale motion to small-scale motion or vice versa. The phenomenon can be observed in a bathtub: large eddies created by stirring result in progressively smaller eddies. A simple argument based on this assumption leads directly to the result that the energy density of intermediate scales of motion in a fluid far from its boundaries varies as the five-thirds power of the scale of motion. This qualitative result is supported by considerable observational evidence. The theoretical treatment does not yield energy levels quantitatively, however, and it includes the reservation that the results apply only to scales of motion so small that they lie outside the "energy bearing" range. This may seem a high price to pay for the "universality" of the five-thirds power result. The hope, however, is that new studies based on similar ideas will reveal statistical aspects of the mass-transporting and energy-bearing range of motions that will depend only weakly on the details of mechanism—what is actually happening in the fluid—and so will more accurately portray the average flow properties.

Discovery Through Computer Simulation

Observation of the shear-flow transition process and its violent nature, described at the beginning of this section, can suggest aspects of fluid mechanism that should be reflected in useful statistical treatments of turbulence. Deeper understanding of this transition has emerged recently from numerical experiments on the growth of small disturbances on shear-flow waves of large amplitude. These studies address only a part of the problem of explaining how the self-propelled wave system works overall. Nevertheless, the results calculated for the wavelength and angle of the initial instabilities (seen on the third bow waves in figure 1) appear to be correct. The region of the transition spot between bow wave and disordered flow has been likened to a hydraulic jump—such as a tidal flood with a high, abrupt front—but here between two waves. The flow of wave energy into the jump region might explain the exceptional rate of growth of the amplitude of the secondary disturbance in the region.

The first rational interpretation of the disordered flow that follows these secondary waves may be found in the similar behavior of the simple systems discussed later in this chapter. Undoubtedly, many other aspects of the transition spot of figure 1 will be reproduced soon in numerical experiments on computers and then dissected theoretically to establish the features of the spot that are important in fully evolved turbulent flow.

Control of Turbulence

An understanding of the mechanism of instability would offer the hope of exercising some degree of control over it. Even though only a little understanding has been attained, a degree of control already has been achieved. The introduction of a few parts per million of certain polymers into a fluid in turbulent shear flow is found to reduce drag dramatically and to modify properties of the flow that were believed to be fundamental. In one experiment, each of several polymers, when added in sufficient amount, produced the same maximum increase in the average velocity of the fluid. The results indicated that the process basically responsible for the increases in velocity depends on a property of the fluid, not of the additives. Further work on modifying instability in turbulent shear flow undoubtedly will be valuable in isolating the elementary processes, on the one hand, and in controlling them for practical purposes, on the other. Drag-reducing additives are being used now for such purposes as reducing pumping costs in petroleum pipelines and increasing the height to which firefighting equipment can propel streams of water.

Thus, although we have statistical models of shear turbulence that are useful for many engineering purposes, further significant progress appears to require a better grasp of the underlying processes. Help may come from additional study of the transition spots involved in the onset of turbulence. Meanwhile, research on the second broad type of disorder, thermal, or convective, turbulence, shows somewhat greater promise of clarifying basic mechanisms.

CONVECTIVE TURBULENCE

In addition to studies of shear turbulence, scientists in the field have long been studying instability leading to turbulence produced by heating a fluid from below. Thermal, or convective, turbulence, like shear turbulence, is a nonlinear process. Initial convective flows, however, are more amenable to mathematical analysis, and theory has yielded excellent pictures of instability. Recent theories of the transition from smooth to disordered flow suggest that, of the two types of turbulence, thermal turbulence may be the more accessible starting point for achieving a general understanding of transitions to turbulence.

Most people intuitively are better able to visualize the nature of fluid motions induced by heat than the complicated processes that result from instability in shearing flow. They are familiar with hot air rising over a radiator or the buoyant plumes that produce a fair-weather cumulus cloud on a summer day. Thus, while some scientists were still struggling for a start in the study of instability in shearing flow, others had theoretically comprehended most types of instability leading to convection.

Theory predicted that a layer of fluid heated from below and cooled from above would develop rising motions in the form of simple, roll-like cells, converting gravitational potential energy into energy of fluid motion. Experimentally, these motions were observed to become disordered only if the difference in temperature across the fluid were increased significantly. It was not observed in these early investigations, however, that disorder could occur in the very first instance of convective instability, which has turned out to be the case.

A Rotating System

An example of disorder in the initial stage of convection appears in figure 2. The six pictures are a time sequence. They were taken from a position above the convecting layer in an experimental system that was being rotated slowly and steadily about a vertical axis (the circle in the center of each picture is an end view of the axis). Each picture was taken at the same angle of rotation, and together they show the development of the horizontal structure of convection. The fluid in the system is methyl alcohol. Its index of refraction changes with temperature; the dark areas in the pictures are hot, rising fluid, and the light areas are cold, descending fluid.

The pictures show that rather smooth, roll-like motions have become unstable in both space and time as a consequence of the rotation of the entire system. This type of disorder has a number of the features expected in a turbulent process, but it is also very different from the wave patterns seen in shearing flow. To the theoretician, the particular virtue of this disordered flow in a rotating system is that it occurs immediately after the mathematically accessible instability leading to the convective motion. Indeed, very recent studies suggest that this type of disorder may be the first to yield fully to theoretical analysis.

Many Disturbances

The first analysis of the initial instability in the simplest type of convective process led to the conclusion that, beyond a critical value of the difference in temperature across the fluid, disturbances of many different sizes could grow and interact. Subsequent study of fluid behavior beyond this critical temperature difference has explained the growing cellular disturbances observed in the ordinary convective process. To study the secondary instabilities of steady cellular convection, certain aspects of the full nonlinear problem must be incorporated into the approximate mathematical description of the instabilities. In a series of

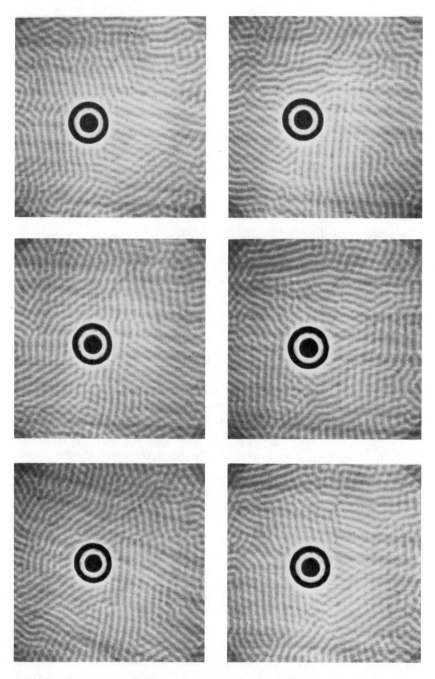

Figure 2   Roll instability in a rotating system. [SOURCE: F. H. Busse and K. E. Heikes. "Convection in a Rotating Layer: A Simple Case of Turbulence," *Science*, Vol. 208, No. 4440 (April 11, 1980), p. 174.]

investigations using such methods, the many types of unstable fluid motions that, under various conditions, can draw energy from the steady convection have been determined. The names of these unstable motions include the oscillatory, the skewed-varicose, the cross-roll, the knot, and the zigzag. Yet, despite the variety of the motions, none leads directly to the onset of turbulence.

The idea that thermal turbulence in a fluid must follow the initial instability immediately was abandoned many years ago. In its place came a view that the first instability would be followed by a second, a third, and many more instabilities as the temperature difference across the fluid layer was increased. As each instability was added to an earlier one, the motion would soon become so complex that the observer would interpret it as turbulence. This view of the origin of thermal turbulence persisted for many years and has considerable experimental support. But, on theoretical grounds, this process would not lead, even after a number of successive instabilities, to the disordered flow with the continuous spectrum of motions observed experimentally.

Only recently has the process leading to thermal turbulence been reassessed. Theoretical inquiry into the nature of unstable systems suggests that if a system experiences a first instability followed by a second instability, a third instability would lead directly to a disordered flow that could be interpreted as turbulent. This view is supported by experimental observations involving measurement of heat flux, or rate of transfer of heat, in a fluid. The flow has taken on a time-dependent and disordered character by the third transition seen in the heat-flux data. Disorder in shear flow, on the other hand, occurs in the abrupt, first shear-flow transition. Hence, more than one path to turbulence evidently exists.

Statistical Approaches

After the first few instabilities that lead to disordered convection, mathematical complexity compels us to resort to statistical theories of turbulence. Several of the theories developed for shearing flow are also applicable to turbulent convection.

The first and simplest use of statistical theory with convection is based on "mean-field" equations. These equations are drawn from the complete equations of motion by considering only those aspects of the fluctuations that modify the fields of average velocities and temperatures in the fluid; the effects of other fluctuating aspects of the flow are discarded. The virtue of the mean-field approach is that it is totally quantitative, so its accuracy can be established easily by comparison with experimental data.

Mean-field calculations turn out to be quite inadequate when the viscosity of the fluid is small, so that instabilities resulting from shearing flow become a factor. When viscosity is large, however, so that the fluid remains relatively free of shear sources of disorder, heat flux predicted by the mean-field calculations appears to agree with the averaged experimental observations within a factor of two. Therefore, despite the huge jump from initial convection to full turbulence, occasionally, the quantitative results of mean-field studies are approximately correct. Mathematical complexity thus far has barred similar studies of shear flow, but such work probably will be done within a year or so.

The quantitative mean-field results for thermal turbulence might be improved by following the closure procedure of the previous section. But a quite different class of turbulence theories, called openers, also offers hope of quantitative understanding. Openers, like closures, consider statistical averages of the full equations of fluid motion. Rather than terminating the sequence of equations by hypothesis, however, the opener procedure is to explore the entire class of possible fluid motions permitted by the first few statistical equations. Among the many possibilities, the solution is sought that provides an upper bound, or maximum value, for some important aspect of the flow. In convection, an upper bound is sought for heat flux. In principle, this calculated upper bound could be brought nearer and nearer to the actual convective heat flux by adding higher order statistical equations to reduce the number of possible fluid motions.

Upper-bound theory has produced the only dependable quantitative results available in all of turbulence theory. The calculated upper bounds—for heat flux, for example—are guaranteed to be higher than the values found experimentally. However, the upper bounds calculated to date are not particularly close to the experimental values. In one case, though, that of turbulent convection with large viscosity, upper-bound theory agrees exactly with the mean-field result for fully evolved convection. Such agreement establishes that the role of the fluid fluctuations neglected in the mean-field approach is to reduce heat flux.

Only a few scientists have worked with upper-bound theory, yet it appears to offer considerable opportunity for use in more complicated flow situations. While conclusions reached with the theory are formally correct, it is possible to improve the accuracy of the calculated upper bound by adding higher order statistical equations to the calculation. In the past, such extensions of upper-bound results have not been feasible but, with the greater availability of computing facilities, upper-bound theory provides a unique way to pin down statistical aspects of turbulent flow in an ordered manner.

Antiturbulence

In the rotating system of figure 2, the convective instability exhibited disorder in its very first appearance. Beyond this initial behavior, increasing the temperature difference across the layer of fluid leads to further disordering. In some recent remarkable experiments well into the turbulent range, however, an instability leading to a persistent, large-scale, ordered flow has been observed.

The experiments were done in a tank full of water with a convective region between two flat plates 10 centimeters apart. Heating the bottom plate produced small-scale, very disordered, convective bursts, but no large-scale motions. Then, unexpectedly, the large-scale, ordered motion appeared and flowed steadily around the tank in the same direction for several days, while the small, disordered bursts continued to appear and decay. Such discoveries could be interpreted as antiturbulence, since the large-scale flow apparently is independent of time but is driven by the very disordered, small-scale convective bursts that come and go in space and time.

Closely Coupled Theory and Experiment

Because of their accessibility to quantitative theoretical study, the early stages of convection under various circumstances have taken on an exciting new role in the growth of our understanding of the advent and development of turbulence in fluids. Nevertheless, thermal turbulence is but one manifestation of the process, and mathematical description of the advent of turbulence is in no sense a description of fully turbulent flow. The goal remains to determine the statistical structure of fully developed flows, whether thermal or shear, both as a philosophical and a practical matter of considerable import.

DISORDER IN SIMPLE SYSTEMS

Arising in part from studies of convection, and related to them through uncomplicated models, is the discovery of simple mechanical and mathematical systems with exact solutions that appear to display disordered behavior. Studies of these systems do not, in fact, lead directly toward the traditional problems of turbulence. But the systems suggest new avenues toward an understanding of disorder in nature, in particular, turbulent processes, and have produced growing excitement among scientists during the past decade.

Many scientists had believed previously that disordered behavior of a mechanical system was a consequence of noise, or extraneous effects

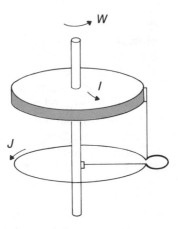

Figure 3 A dynamo (*W* = rotation rate of metal disk, *I* = electric current in disk, *J* = current in stationary wire coil). [SOURCE: K. A. Robbins. "A Moment Equation Description of Magnetic Reversals in the Earth," *Proceedings of the National Academy of Sciences*, Vol. 73, No. 12 (December 1976), p. 4298.]

caused by external disturbances. It might be thermodynamic noise or even the effects of trucks shaking the ground in the distance. New observations, however, show that elementary processes in certain systems lead to exactly reproducible disorder. Such systems are said to contain a "strange attractor"—a feature that attracts the system toward disordered behavior. The disorder is time-dependent—it is not also a function of position as would be the case with velocity and temperature in a fluid. It has come to be called temporal chaos. The hope is that generalizable features of this temporal chaos will shed new light on the fully developed disorder in both space and time that is characteristic of shear and thermal turbulence far from their onset.

Disorder in a Dynamo

An example of a system that exhibits reproducible disorder is the homopolar dynamo (figure 3), a simple form of electrical generator. A dynamo converts mechanical energy to electrical energy. The shaft of the metal disk in figure 3 is driven by a steady torque. When the disk reaches a certain rotational velocity (*W*), infinitesimal fluctuations of current (*J*) in the wire loop below the disk produce a magnetic field which, in turn, is reinforced by currents (*I*) flowing in the disk because of its motion through that magnetic field. This system can exhibit magnetic instabilities of either polarity—that is, in the magnetic field outside the system, "north" can be either up or down.

It has been known for several years that this dynamo can behave

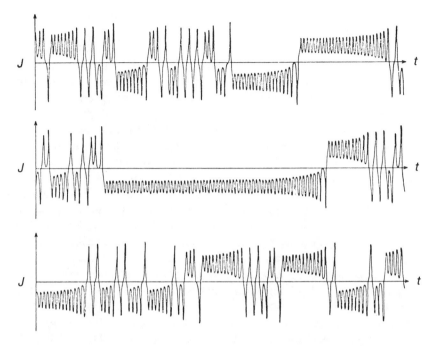

Figure 4 This figure is a record, over time (t), of the current (J) flowing in the wire loop of the dynamo in figure 3, when the load on the dynamo is purely resistive. The record suggests that the polarity of the magnetic field reverses at random intervals, yet the behavior can be reproduced exactly, time after time, from the equations describing the behavior of the dynamo. What seems random, in other words, is exactly reproducible disorder. [SOURCE: K. A. Robbins. "A Moment Equation Description of Magnetic Reversals in the Earth," *Proceedings of the National Academy of Sciences*, Vol. 73, No. 12 (December 1976), p. 4299.]

very strangely in time. If the small external loop on the lower right, which represents the load on the dynamo, is purely resistive, the system becomes unstable to growing oscillations. Figure 4 is a record of the current that flows in the loop as a function of time. These data suggest that the polarity of the magnetic field reverses at apparently random intervals. Yet, the behavior is not a consequence of extraneous external disturbances; it can be reproduced exactly time after time from the equations describing the behavior of the dynamo. The reversals of polarity are most nonuniform at the very lowest torque at which the instability occurs; this process has been suggested as an explanation of the magnetic reversals that occur in the geodynamo deep in the liquid core of the earth, causing the north and south magnetic poles to change positions as often as every 15,000 years.

The significant feature of this disordered behavior is that it can occur in a system with only three degrees of freedom, or variables: the angular velocity of the dynamo's shaft, the current in the disk, and the current in the wire loop. In other words, the system is completely accessible for detailed study, in part by mathematical analysis and in part with computer assistance. Such accessibility raises the hope that an understanding of this type of disorder, once achieved, may help to clarify the much more complicated disorder of fluid systems in space and time.

Equations that describe the homopolar dynamo are vastly simpler than those that describe a fluid system. The variables in the dynamo equation, as noted above, are functions only of time and not also of position, as are the velocity and temperature fields in a fluid system. The dynamo equations, however, also describe the first beginnings of the convective process after the point of instability. These are called the Lorenz equations.* Exploration of many equations has not revealed any simpler than the Lorenz equations which describe a continuous physical process that can be set up in the laboratory.

In these equations, a, b, and c represent the time-dependent variables, W, I, and J, of the homopolar dynamo. The dot that appears over the first three symbols ($\dot{a}$, $\dot{b}$, and $\dot{c}$) represents the rate of change of those quantities with time. When the rates are zero, the three equations are three algebraic relations that determine the steady-state solutions of a, b, and c. R is the experimentally controlled constant, determined in the convective case by how much heat flows across the fluid layer, and in the dynamo case by the amount of torque applied to the shaft of the dynamo. P, in the convective case, is the ratio of thermal to viscous diffusion coefficients; in the dynamo case, P is the ratio of the two time constants in the electrical circuit.

Approaches to Chaos

Solutions of the Lorenz equations describing the behavior of the dynamo reach a chaotic state by an abrupt transition. This type of transition from order to disorder is akin to the abrupt transition to disorder observed in fluids in shearing flow.

Simple systems also may approach chaos through a sequence of period doublings of an oscillatory part of the motion. An example of such behavior was found in experiments with a small convective cell containing liquid helium. The cell was heated at the bottom and the

---

*Lorenz equations:

$$\dot{a} + a - bc = 0$$
$$\dot{b} + b + ac = R$$
$$P\dot{c} + c - a = 0$$

flow of heat at a point in the helium was measured. When the temperature difference across the cell reached a critical value, the heat flow began to oscillate with a time period of t. At a second, larger temperature difference across the cell, the heat flow did not repeat after a time t, but after $2t$. As the temperature difference was increased, the period of oscillation continued to double until finally the motion became chaotic.

This sequence of period doublings appears to have generalizable aspects far beyond the problem of turbulence. The values of the parameters, such as temperature, at which the period doublings occur lead to a ratio that appears not only in these experimental data, but in purely mathematical studies (such as mapping of the unit interval).

Persistent Order

Despite the temporal chaos seen in simple systems, a degree of order persists. When the current that flows in the loop of the homopolar dynamo, for example, is plotted against time, it appears to fluctuate, or change direction, quite randomly. When the sequence of maximum currents at each fluctuation is plotted, however, the result is an ordered cusped curve. A significant feature of such plots is that they are not altered by small disturbances, or noise. A small amount of noise imposed on the homopolar dynamo would change the plot of current fluctuations with time to the point where it would bear no apparent relation to the original plot. But the resulting cusped curve—the plot of maximum currents—would show no change from the original, even if the noise were imposed continuously. Such plots, therefore, show not only an order inherent in the temporal chaos but a stability of that order in the presence of imposed disturbances.

Other properties of these simple systems also show interesting order amid the chaos. Examples are the heat flow in convection and the average strength of the magnetic field in the dynamo. Both processes speed up in the same fashion with the addition of greater forcing, such as increasing the temperature difference across the convection. Explorations of this correlation and the reason that the dynamo and the simple convection retain these qualities of the steady solutions to the equations have yet to be made. Indeed, many aspects of temporal chaos remain mysteries. They are mathematically accessible, however, and should be clarified in the near future.

These transitions to temporal chaos have been compared to thermodynamic phase transitions from a more ordered to a less ordered state. The differences are considerable; no underlying theory connects the two processes. The discovery of a mathematical method called renormalization group theory, however, has provided a powerful tool for interpreting thermodynamic aspects of phase transitions in condensed matter (generally, the solid state). This theory makes it possible to identify new, universal features of physical processes, including aspects of turbulence. One example is the period doubling ratio mentioned earlier.

Application of the theory to the statistically steady states of simple dynamic systems has proved valuable in some instances, but not yet in others. Renormalization schemes, for example, can be used to address the more ordered behavior, such as the period-doubling process, preceding the chaotic state. Yet, limited success has attended applications of renormalization group theory to the traditional problems of turbulence.

A search is in progress for representational frameworks or special problems in which this new technique can capture statistical features of fully evolved turbulence in fluids. A significant guide gained from study of the dynamo problem is that the chaotic state that is realized, contrary to what might have been expected, can be shown mathematically not to be the one that would result in the highest rate of dissipation of energy.

It is difficult to predict where the current studies of simple systems will lead, but they are beginning to make it possible to isolate the basic elements responsible for disorder. The aspects of simple systems that are currently accessible to analysis do not lie at the heart of the traditional problems of fully evolved turbulence. Nevertheless, investigations of disordered behavior in simple dynamic systems promise to contribute substantially to new understanding of the turbulent process.

Outlook

The discovery in recent years of mechanical and exactly solvable mathematical systems that lead to chaotic behavior suggests new paths toward understanding turbulence in fluids, whether gases or liquids. Turbulence is the cause of bumpy airplane rides, the rapid mixing of chemicals in industrial processes, the enhanced drag on petroleum flowing in pipelines, and many other phenomena that can be either desirable or undesirable. Scientists and engineers concerned with turbulence are beginning to think more about the mechanism involved, as opposed to its statistical manifestations—more about what actually happens in the fluid, as opposed to statistical approximations of what happens. Such work is yielding a new scientific language—a new way of thinking about turbulence—that promises deeper understanding in the search for universally applicable features of the turbulent process.

Statistical approximations, based on experimental data, are necessary because we have no other way of dealing with turbulence in engineering design and for other practical purposes. The theory of this immensely com-

Fighting turbulence. [Fish and Wildlife Service, U. S. Department of the Interior.]

plex process, after more than 100 years of scientific study, has yielded negligibly few quantitative predictions of turbulence. The trend toward mechanistic thinking will not, soon, bring complete or even close to complete understanding of the central problem, fully evolved turbulence. But it offers every hope of leading to the construction of statistical hypotheses that reflect more accurately the underlying mechanism of the process.

For the near and intermediate term, the goals are to increase our skill at controlling instabilities and to achieve a broad predictive capability for statistical models of turbulence. Such capability would permit both energy saving and improved performance in redesigned pumping systems, compressors, turbines, aircraft, chemical reactors and mixers, and many other products. New understanding of the limits on predictability of the large-scale disorder in weather and in ocean currents can be an early reward of the studies of chaos in simpler systems. New awareness of the origins of disorder in natural processes may be as important a contribution to model building in economics, political science, and ecology as it is in fluid turbulence.

BIBLIOGRAPHY

F. H. Busse. "The Optimum Theory of Turbulence," *Advances in Applied Mechanics*, Vol. 18 (1978), pp. 77–121.

O. E. Lanford, III. "The Strange Attractor Theory of Turbulence," *Annual Review of Fluid Mechanics*, Vol. 14 (1982), pp. 347–364.

S. A. Orszag. "The Statistical Theory of Turbulence," in *Fluid Dynamics*. Edited by R. Balian et al. Paris and New York: Gordon & Breach Science Pubs., Inc., 1977.

7

Lasers

In the brief quarter century since the laser was invented, intensive research and development have made it an essential tool in science and technology. The device has led to breakthroughs in our fundamental understanding of the interaction of radiation with matter and is permitting us to measure the most fundamental physical quantities. Lasers are used in medicine, industry, and defense. Progress in the field has been remarkable by any standard. But fully as remarkable are the great promise and incredible vitality that continue to characterize laser science and technology.

LASERS AND COHERENT LIGHT

By now, laser is a household word, but it began life as an acronym for "light amplification by stimulated emission of radiation." The device employs high-intensity light or electric current to cause atoms, molecules, or ions in a medium to radiate at frequencies corresponding to the separation between discrete energy levels. This process generates light of essentially a single wavelength, called coherent light. The wavelength varies with the nature of the light-emitting, or lasing, medium selected; the medium may be solid, liquid, or gaseous—for instance, ruby, various organic dyes, and carbon dioxide. Some lasers can be tuned over a range of wavelengths; lasers can be designed to generate light in pulses or continuously (continuous wave, or cw).

Light from a laser is coherent for all practical purposes. That is, it covers a relatively narrow range of wavelengths called the bandwidth or linewidth. Usually, the bandwidth is given in hertz (Hz), a measure of frequency. Laser bandwidths typically are on the order of 10 megahertz, although bandwidths of a few hertz have been achieved experimentally.

Coherent light has unique properties: it is monochromatic (of one color); it is highly directional, emerging from the laser as a narrow, focused beam; and it has high power. Efforts to exploit this combination of properties have led to a variety of lasers and numerous applications, many of them impossible with other techniques or equipment.

Besides its immense scientific utility, the laser is the basis of lightwave communications systems operating throughout the world. A laser-based process for separating isotopes of uranium is scheduled for major scale-up in this country. Lasers are used commonly in such manufacturing processes as drilling, cutting, welding, and surface treatment; they are readily adaptable to computer-controlled fabrication. Defense technology benefits from such equipment as laser radar and laser target designators. Laser techniques are establishing new frontiers in medical diagnoses and surgery.

The prospects in laser science and technology treated here were selected because of their potential importance during the next five years. They fall under five general headings: new lasers and laser systems,* high-resolution spectroscopy, lightwave communications, separation of isotopes, and applications in industry, defense, and medicine.

NEW LASERS AND LASER SYSTEMS

In 1982, existing lasers satisfy the needs of science and technology over much of the electromagnetic spectrum. In the infrared (long-wavelength) region, there are carbon-dioxide and other molecular lasers. In the near-infrared region, there are neodymium-doped glass, neodymium-doped yttrium aluminum garnet (YAG), and other lasers. Several types of ion lasers are used in the visible region.

Still other laser systems are needed urgently, however, to provide coherent light with specific characteristics. For example, there is no laser that produces light in the visible range with relatively high power. Prospective uses of such a laser would include large-screen displays and photochemistry. Laser sources become scarce as one approaches

---

*The laser or the laser source is the lasing medium, and the laser system is the medium plus the associated equipment. Laser will be used here to mean both, except where the distinction is important.

the near-ultraviolet and shorter wavelengths. Needs exist also for coherent light of shorter pulse durations, or times, than are now available. Prospective uses for shorter wavelengths and pulse durations are particularly evident in basic scientific investigations and measurements.

Most measurements in the physics of solids, liquids, and gases are designed to determine either structure (geometry) or kinetics (the rate of chemical or physical change). The laser is a powerful tool for high-precision structural studies and, during the past 20 years, it has been the major tool in kinetic studies because it can provide either uniquely high spectral resolution or pulses of uniquely short duration. Many chemical reactions, for example, occur in times on the order of a picosecond (ps—10^{-12} second); spectroscopic study of reactions, using lasers with pulse durations approaching a picosecond, yields unprecedented insight into chemical processes. The laser has become so central in kinetic studies that the advent of new spectral ranges or shorter pulse durations has uncovered entirely new directions of research.

Shorter Pulses, Shorter Wavelengths

Until recently, advances in lasers have been confined largely to pulses longer than a picosecond in the wavelength region of 0.2 to 2.0 micrometers (μm—10^{-6} meter); these wavelengths are in or closely surrounding the visible region. Lately, however, it has become possible to produce pulses as short as 50 femtoseconds (fs—5×10^{-14} second) and wavelengths as short as 0.05 μm. Lasers with these characteristics expand the range of investigation to such phenomena as very fast chemical reactions; preferential excitation of isotopes, which in part can be used to separate them; and changes in the electronic state of solids such as those used in electronic devices.

Coherent light with wavelengths shorter than about 200 nanometers (nm) has been available until recently only in specific regions of the spectrum and with specific characteristics. It has been produced, by and large, by nonlinear optical techniques, in which light that enters a medium at a given wavelength emerges at a different wavelength (in a linear interaction, it emerges at the entering wavelength). Specifically, light of an appropriate longer wavelength from a powerful source is converted by the nonlinear medium into harmonic frequencies by processes called harmonic generation and sum-frequency mixing to give the desired shorter wavelength.

Recent research has identified gaseous media in which the sum frequency of three visible lasers is generated quite efficiently. As a result, it is now possible to produce, rather routinely, tunable coherent light at any wavelength between 200 and 105 nm. This makes it possible, for example, to measure the critical hydrogen content in the plasmas of tokamaks, reactors used to study nuclear fusion. It also makes possible the measurement of the concentrations of fluorine and chlorine in reactive-ion etching vessels, where those elements are used to etch electronic circuits on silicon chips.

The power output from sum-frequency generation processes is in the range of tens of microwatts, which is more than adequate for investigating linear interactions but not for carrying out further nonlinear interactions. Thus, work is continuing on the development of true laser sources in the vacuum ultraviolet (short-wavelength) region. Recently, a dissociation Raman laser using sodium iodide as a medium and generating coherent light at 178 nm was demonstrated successfully. In this device, broadband light dissociates the sodium iodide into its constituent atoms in their excited states. The resulting medium, upon application of laser light in the visible range, produces the short-wavelength light through the mechanism of stimulated anti-Stokes Raman Scattering.

Production of coherent light at wavelengths shorter than 105 nm (in the extreme vacuum ultraviolet) requires special technologies because in that range all window materials become "lossy"; that is, light cannot enter or leave the laser. During the past year, efforts to circumvent this problem have led to a sum-frequency source, based on mercury vapor, which is tunable from 120 to shorter than 80 nm. Research is expanding rapidly, and tunable, sum-frequency sources should be able to cover the spectrum to wavelengths as short as 50 nm within the next few years.

What about wavelengths shorter than 50 nm? Sum-frequency techniques are not likely to produce useful quantities of light in that region of the spectrum until powerful primary lasers in the vacuum ultraviolet are developed to provide the necessary longer wavelength light. A need exists, however, for sources of pulsed, high-intensity light at very short wavelengths, down to and including soft X-rays (wavelengths of less than 5 nm). Uses of such light would include time-resolved photoemission for studying energy levels in solids, and also work in microscopy requiring very small focal spots with sizes determined by the wavelength of the incident light. Other potential uses include holography and the study of biological systems by diffraction and a method called extended X-ray absorption fine structure (EXAFS). To this end, vigorous efforts are under way to produce extremely bright sources of incoherent light using laser-induced plasmas (figure 1). Such plasmas are promising sources for optical pumping systems for vacuum ultraviolet lasers.

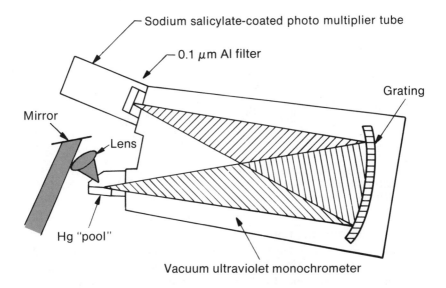

Sodium salicylate-coated photo multiplier tube

0.1 μm Al filter

Grating

Mirror

Lens

Hg "pool"

Vacuum ultraviolet monochrometer

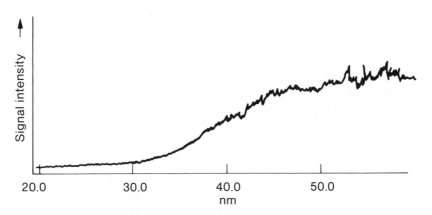

Figure 1 Apparatus (above) used to produce incoherent light from a laser-induced plasma in mercury (Hg). The continuum emission recorded from laser-induced plasma is shown below. [SOURCE: Bell Laboratories.]

Ultrashort Pulses

A technique called mode-locking is now used routinely in the laboratory as well as commercially to produce pulses of laser light with durations of about 1 ps. Recently, a new method involving colliding pulses in a ring laser cavity has yielded pulse durations as short as 30 fs; this extremely short pulse is then amplified to peak powers of 2×10^{10} watts (W).

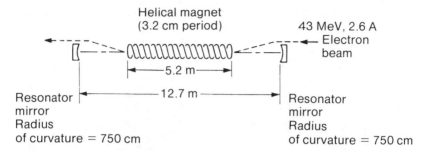

Figure 2 Schematic of a free-electron laser. [SOURCE: D. A. G. Deacon et al. "First Operation of a Free-Electron Laser," *Physical Review Letters,* Vol. 38, No. 16 (April 18, 1977), pp. 892–893.]

A pulse duration of 30 fs is shorter than the time it takes for two atoms in heavy molecules to vibrate, for example, or for an atom to change sites on a surface. This unparalleled time resolution has spurred research into the very fundamentals of collision physics—the processes that occur when atoms and molecules collide. In addition, the extremely high peak powers of these very short pulses have permitted investigators to generate (via nonlinear processes) pulses of light with a duration of less than 0.1 ps and a spectral width covering the infrared, visible, and ultraviolet regions of the spectrum. This continuum will yield the equivalent of a snapshot of a wide range of molecular processes—on a surface, for example—at one point in time with a resolution of better than 100 fs.

Free-Electron Lasers

Breakthroughs in the technology of tunable laser sources have led to significant gains in the study of new phenomena and materials by optical spectroscopy. At wavelengths shorter than 250 nm or longer than 25 μm, however, the existing laser and nonlaser sources of light lack either the versatility or the flexibility needed to explore new science. In addition to the progress already noted with short-wavelength sources, a recent development—the free-electron laser—promises to deliver tunable coherent radiation at both the short- and long-wavelength ends of the spectrum. In principle, it could be done with one laser system, but it probably would not be done that way in practice.

In a free-electron laser, a beam of relativistic electrons (traveling at nearly the speed of light) is injected into a wiggler magnet, a periodically alternating magnetic field (figure 2). Amplified radiation is produced at a wavelength determined by the energy of the beam, the spatial periodicity of the magnet, and the strength of the magnetic field.

The device possesses the unique advantage of broad tunability. Compared to other types of lasers, it can be tuned relatively easily by changing the energy of the electron beam.

As yet, free-electron lasers have produced coherent radiation only at wavelengths of 3.4 and 400 μm and 2 millimeters, and no tunable laser has been demonstrated. Studies of the devices at Stanford University, Columbia University, and the Naval Research Laboratory, however, have confirmed the general nature of their tunability. Operation of free-electron lasers as sources of tunable, coherent, high-power (both cw and pulsed) radiation in the far infrared (long-wavelength) and vacuum ultraviolet (short-wavelength) regions is anticipated.

Besides the free-electron lasers at Stanford University, Columbia University, and the Naval Research Laboratory, at least four new ones are near completion. These are the project at Santa Barbara, using a Van de Graaf accelerator as a source of relativistic electrons; the Los Alamos project, using a radio-frequency linac; the Brookhaven project, using an electron storage ring; and the Bell Laboratories project, using a microtron. The Bell Laboratories project is the only one specifically designed to provide tunable radiation for spectroscopic applications in the far infrared—a region of the spectrum rich in new phenomena to be explored in the area of solid-state physics (see figure 3).

HIGH-RESOLUTION SPECTROSCOPY

High-resolution laser spectroscopy is developing rapidly with respect to laser sources, techniques, and applications. The major areas of interest are spectroscopy of simple, fundamental systems; spectroscopy of more complicated atomic or molecular systems; applications in solids; and ultraprecise stabilization of lasers and the associated metrology (the science of measurement).

Fundamental Systems

High-resolution spectroscopy is being used on fundamental systems where the basic physical theory is believed to be known and, in principle, can be applied with very high accuracy. The spectroscopic techniques yield precision measurements that can be used to test the theory and calculations or to determine the values of fundamental constants. In general, this approach is designed to test and refine the ideas of quantum electrodynamics.

The prime example is the detailed study of the hydrogen atom or hydrogenlike ions, which are so simple that their fundamental characteristics, such as energy levels, can be described very accurately by quantum electrodynamics. Separation between energy levels can be

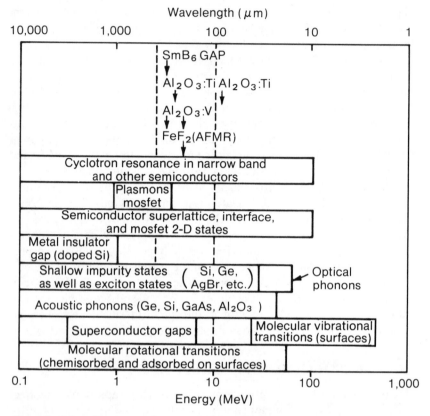

Figure 3 Atlas of important elementary excitations in solids covering a wavelength region from 1 to 10,000 μm. [SOURCE: Bell Laboratories.]

measured very precisely with lasers. One such separation (the 2S–3P interval) was measured recently to an absolute accuracy of one part in 10^9 using a crossed atomic/laser beam; this accuracy will be extended to a few parts in 10^{10} in the near future.

Advances also have been made with hydrogenlike ions. In one such ion, a form of chlorine (Cl^{16+}), an energy spacing (the 2S–2P interval) has been measured to an accuracy of about 1 percent using a carbon dioxide laser. The measurement cannot be classified as a high-resolution determination in a strict sense, but the results nevertheless provide a crucial check on the various theoretical calculations.

One of the most fundamental systems is the positronium atom, the bound state of an electron and its antiparticle. Recent success in optically exciting this two-body system has raised the possibility of studying it in detail. The 1S–2S interval in positronium has been measured to an accuracy of one part in 10^6 using a high-power, narrow-bandwidth, tunable dye laser. The accuracy will be improved by two orders of magnitude in the near future.

Complicated Atoms and Molecules

High-resolution laser spectroscopy of more complicated atomic and molecular systems is being actively pursued and finds many applications. These systems can serve as proving grounds for many spectroscopic techniques. For example, the two-photon, Doppler-free, absorption technique, a method of exciting atoms for high-resolution spectroscopic study, was first demonstrated in sodium atoms. High-resolution spectroscopy of molecules and molecular ions is used also to study the earth's upper atmosphere and interstellar and intergalactic media.

In addition, high-resolution spectroscopy using lasers can detect the very small, parity-nonconserving effects that result from weak interactions, or forces between uncharged bodies. Recent studies of bismuth, thallium, and cesium using a tunable dye laser have verified a theoretical model (the Weinberg–Salam model) unifying weak and electromagnetic interactions. Such results, again, build confidence in the quantum electrodynamics theory.

Solid Materials

Much information on the internal structure and behavior of solid materials is coming from high-resolution laser spectroscopy. A novel, high-resolution technique, for example, can measure directly the resonant transfer of energy among chromium atoms in ruby. High-resolution light-scattering can be used to clarify changes in structural phases in crystals, polymers, and amorphous materials. By steadily expanding the knowledge base of solid materials, such data contribute to our ability to exploit these materials in electronic or other kinds of devices.

A new way to store information could result from the finding that decay of energy levels in a solid composed of lanthanum fluoride and praseodymium yields linewidths as narrow as 2 kilohertz. These linewidths are called homogeneous because they arise from the intrinsic lifetimes of energy levels; aggregates of different homogeneous linewidths centered at different frequencies result in inhomogeneous lineshapes. The existence of narrow homogeneous linewidths within relatively broad inhomogeneous linewidths has potential for optical storage of information at ultrahigh density, since each small physical area of memory can contain many possible logic states. Exploration of this possibility is under way.

A technique developed over the past two years has made it possible to measure absorption of optical radiation by solids that are extremely transparent or, in other words, only weakly absorbing. The method is PULPIT optoacoustic spectroscopy (PULPIT is the acronym for pulsed laser piezoelectric transducer). This technique in part permits studies involving the vibrations of molecules in solid hydrogen. PULPIT optoacoustic spectroscopy has yielded the first observations of the second and third overtone vibrational absorptions in solid hydrogen. Overtone absorption spectra of hydrogen hold considerable interest in interpreting the spectra of the atmospheres of the giant outer planets; these spectra may be important in determining the compositions of those atmospheres.

PULPIT optoacoustic spectroscopy also may fill a longstanding practical need to measure ultrasmall optical absorptions in solids of technological importance. Such absorptions provide a measure of ultrasmall amounts of undesirable impurities. One hoped-for use is with the starting material for fabricating low-loss optical fibers for lightwave communications. Another is with the semiconducting materials used in the optoelectronic technology for studying absorption arising from deep-lying uncharged impurities.

Ultraprecise Stabilization and Metrology

Spectroscopy and metrology share a need for precision, and laser spectroscopy promises to be the basis for new and more precise standards of time and length. Recent progress in the ultraprecise stabilization of tunable dye lasers and the associated metrology is truly remarkable. Linewidths of less than 100 Hz have been achieved, and improvements are on the way. Thus, in principle, the resolving power of lasers can surpass that of the most precise radio-frequency or Mossbauer techniques. Before these lasers can be used in spectroscopy, however, ways must be developed to eliminate such effects as those arising from transit time—the time a moving atom can stay inside a laser beam.

Such elements as titanium, indium, gallium, and aluminum have energy levels with lifetimes so long that laser spectroscopy should be able to resolve atomic spectral lines to an accuracy of 1 part in 10^{18}. The prospect of such extraordinary resolution is mind-boggling. The precise clock thus provided should permit ultraprecise experiments in such areas as general relativity, weak interactions, and similar exotic branches of physics.

LASERS FOR LIGHTWAVE COMMUNICATIONS

Developments in terrestrial lightwave communications—the transmission of information via laser light traveling in small optical fibers—are happening very rapidly. (Lightwave communications in space, as between satellites, would not, of course, involve fibers.) Such systems can carry very large amounts of data and offer an alternative—depending on the particular set of economics involved—to transmission by wire or microwaves. Numerous lightwave communications systems are operational throughout the world, and an optical transatlantic cable may be installed linking this country with England within five years.

The past decade has seen laser communications technology spring from its infancy, with the first semiconductor laser capable of continuous-wave operation at room temperature, to the demonstration in 1977 of a semiconductor laser having an estimated operating life of more than a million hours, based on extrapolations from aging tests at elevated temperatures. Of the many lasers available for terrestrial communications, semiconductor injection lasers are used most because of their compatibility with the optical fibers, fabricated from fused silica, that are used to transmit the optical signals.

Advances in lasers and fibers have gone hand in hand, first with improvements in the loss characteristics of the fibers at the wavelengths of the earlier semiconductor injection lasers, and then with the development of lasers at the wavelengths where the optical fibers have even lower losses and minimum dispersion. This section does not concern optical fibers directly, but it should be mentioned that they have reached the point of exhibiting transmission losses of less than 1 decibel per kilometer (dB/km) in the near infrared region of the spectrum. Fibers are anticipated that will have losses orders of magnitude lower than those currently achieved if material purity problems can be solved.

Current Technology

All semiconductor lasers used for communications systems are based on a double heterostructure concept shown in figure 4. The lasing material is in the form of a thin (0.1 to 0.2 μm) layer and is sandwiched between two cladding layers. The cladding material, having a bandgap larger than that of the lasing region, is instrumental in optical and carrier confinement, which lowers the minimum current required to achieve the lasing threshold. In the wavelength range of 0.82 to 0.88 μm, both the active layer and cladding are composed of aluminum, gallium, and arsenic (AlGaAs), but in different atomic proportions. Between 1.2 and 1.65 μm, the active layer is composed of indium, gallium, arsenic, and phosphorus (InGaAsP), and the cladding layers are indium and phosphorus (InP).

Lightwave communications systems now operating are based primarily on the AlGaAs technology. This original technology, however, has been overshadowed already by the second-generation lasers based on InGaAsP. Systems using the latter give better performance because

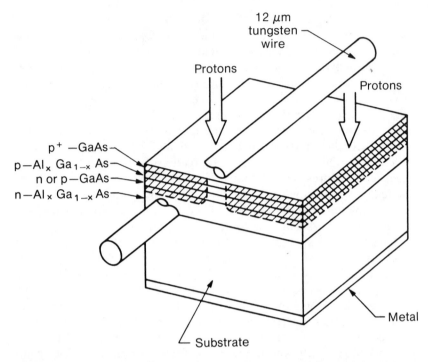

Figure 4   Schematic of a stripe-geometry double heterostructure laser fabricated by proton bombardment. In the first layer, "p+" indicates a very highly doped p-type. [SOURCE: Bell Laboratories.]

the lasers are optimally matched to the transmission characteristics (optical loss and dispersion) of the optical fibers.

The double heterostructure lasers used in commercial lightwave communications systems are relatively easy to fabricate and have been tested extensively for reliability. The light they emit has a relatively broad bandwidth and is transmitted by what are called multimode fibers. More complex and costly laser designs operating in a single mode—that is, at a single wavelength—will employ single-mode fibers. These fibers are 5 to 10 μm in diameter, as opposed to about 20 μm or larger for multimode fibers.

Choice of laser wavelength is dictated largely by the properties of the optical fiber. Current fibers based on fused silica have their lowest transmission losses in three wavelength windows centered at about 0.85, 1.3, and 1.55 μm. A typical loss spectrum of a single-mode optical fiber is shown in figure 5. The lowest loss is about 0.2 dB/km at 1.55 μm. This characteristic of the fibers strongly favors the second generation lasers operating in the low-loss windows centered at 1.3 and 1.55 μm.

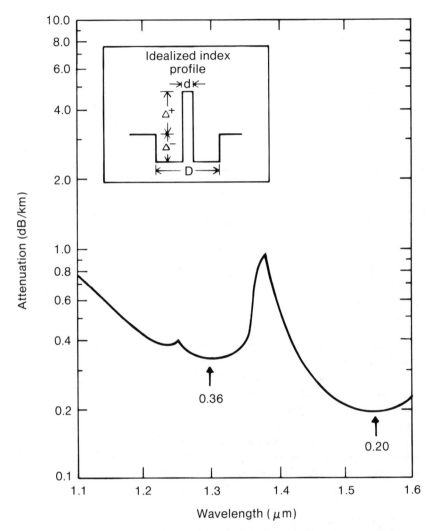

Figure 5 Absorption characteristics of a typical single-mode fiber. [SOURCE: Bell Laboratories.]

The potential of emerging, longer wavelength technology is best illustrated by a recent test of a prototype system employing a 1.3-μm single-mode laser. Information was transmitted at 274 megabits per second over 101 kilometers of fiber at essentially zero error rate and with no repeaters. (Repeaters normally would be located at specified intervals along a fiber-optic transmission line to boost the signal by converting it from optical to electrical form and then, via laser, back to optical form.) The prototype test represents a repeater spacing nearly

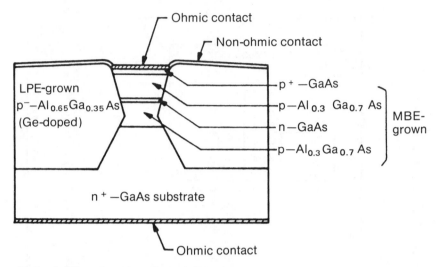

Figure 6   Mirror-face view of a buried heterostructure $Al_xGa_{1-x}As$ laser capable of single-mode operation. LPE-grown $Al_{0.65}Ga_{0.35}As$ is lightly doped with germanium (Ge). [SOURCE: Bell Laboratories.]

an order of magnitude greater than that possible using the shorter-wavelength AlGaAs systems.

Crystals for today's commercially available lasers are grown by a method called liquid phase epitaxy (LPE). The method is relatively simple, but often it produces less uniform epitaxial layers—layers of similarly oriented crystalline structure—than do the newer techniques of molecular beam epitaxy (MBE) and metal organic chemical vapor deposition (see chapter 5). These newer technologies will play increasingly important roles in the fabrication of lasers.

Basic Understanding

The availability of sophisticated, single-mode, semiconductor lasers has made possible several recent studies of their fundamental properties. The buried heterostructure laser shown in figure 6 is ideal for such studies. It is formed by growing a planar, double-heterostructure, semiconductor wafer, etching narrow, photographically defined mesas on it, and burying the mesas by a second liquid phase epitaxy regrowth step. Several variations of the basic design have been demonstrated. By studying the basic gain and loss mechanisms in single-mode AlGaAs lasers, it has been possible to understand the origin of the fundamental linewidth of semiconductor lasers.

Defect reactions that may limit the reliability of semiconductor lasers are important but less than completely understood problems. Reliable lasers can be fabricated but, to develop a reproducible technology, more must be known about degradation mechanisms. At the microscopic level, for example, it is not known whether the observed growth of nonlasing regions within the overall lasing region is due to the creation of defects or to the motion and agglomeration of preexisting defects.

The basic science of ohmic contacts also is incompletely understood. Ohmic contacts are the points at which electrical power enters the laser, and good contacts are essential to the reliability of semiconductor lasers. Yet the technology of ohmic contacts today is more an art than a science. Improved understanding of the interaction of metals with semiconductors is important for ohmic contacts and also to clarification of degradation phenomena.

Future Directions of Technology

Development of communications lasers in the next five years will be focused primarily on InGaAsP/InP systems for the 1.3- and 1.55-μm regions. Major effort will be made to improve the manufacturing yield of such devices. Technological progress that will reduce the cost of lightwave communications systems is likely in three areas: improved growth of materials, wavelength division multiplexing, and higher data transmission rates.

The promising new growth technologies, molecular beam epitaxy and metal organic chemical vapor deposition, provide the uniformity and homogeneity essential to manufacturing control of laser materials. These technologies, especially molecular beam epitaxy, also provide an atom-by-atom control of composition that permits the design of lasers impossible to fabricate by the current liquid phase epitaxy process. In one such design, utilizing a multiple quantum well structure, optical gain is enhanced and the laser threshold current reduced by a factor of 50 over that possible in a conventional design. Wavelength division multiplexing also will enhance the economic attraction of lightwave communications by increasing the amount of information that can be handled on a single fiber. In this technique, several laser channels—currently two to four—are combined on the same optical fiber. The several beams of different wavelengths are mixed at the transmitting end before they enter the fiber and, at the receiving end, the different wavelengths are separated prior to their detection.

An important future direction is very rapid modulation of the laser to give data transmission rates in excess of 10^9 bits per second. Current communications lasers are modulated at about 10^8 bits per second but, at faster rates, the stability of the lasing mode, or wavelength structure, becomes a problem. Thus, new laser designs that permit faster modulation while preserving the mode structure are imperative.

Future Directions of Research

To support technological needs beyond five years, research on communications lasers is likely to focus on such areas as longer wavelength materials, integration of various optical components (integrated optics), ultrafast modulation, and ultrastable lasers, as well as on radically new concepts of laser design.

The need for monolithic integration of lasers with other optical components is most pressing in single-mode optical systems, where optical alignment on micrometer scales is necessary. In five years, we expect to see significant progress in integrating lasers with other elements of laser communications systems. Such integration also would aid ultra-high-speed modulation. In spite of many unknowns, the science of communications lasers will advance more rapidly than we can foresee and will drive the design of lightwave communications systems to benefit from what is possible.

SEPARATION OF ISOTOPES BY LASERS

The principal effort in the use of lasers to separate isotopes has centered on the enrichment of the fissionable isotope of uranium, uranium-235. The concentration of this isotope in naturally occurring uranium is only 0.72 percent, compared with 99.27 percent for the fission-stable isotope, uranium-238. Much higher concentrations are needed for use in nuclear reactors and for nuclear weapons. Although this discussion is concerned primarily with the enrichment of uranium, the techniques are similar to those used to enrich isotopes of other elements.

Traditionally, uranium-235 has been enriched by the gaseous diffusion method. Because of the difference in mass of the two isotopes, molecular gases (uranium hexafluoride) containing them diffuse through a porous barrier at slightly different rates. The diffusion process must be repeated many times, at very high cost, to increase the enrichment gradually to the required level. Laser isotope enrichment, on the other hand, offers the prospect of achieving that level in essentially one step.

The basic idea is to use a laser to excite selectively the internal energy states of uranium-235 atoms or molecules in a vaporized mixture of species of that isotope and uranium-238. This process is possible because, in both atomic uranium and in uranium hexafluoride, the energy levels of the different isotopes are far enough apart so that one will absorb radiation of a particular wavelength while the other will not. Thus, when a vapor or gas containing a mixture of the isotopes is irradiated with a narrow-bandwidth laser beam of the proper wavelength, one isotope absorbs energy from the beam while the other is undisturbed. Then, the excited isotope is extracted selectively from the medium by either dissociation into different species in the case of molecules or ionization in the case of atoms. The narrow bandwidth and high power of the laser beam make this selectivity feasible.

Laser isotope enrichment in the United States has involved both atomic and molecular processes. The Exxon Corporation and Lawrence Livermore National Laboratory have concentrated on the atomic process and Los Alamos National Laboratory on the molecular technique. In 1981, Exxon decided to terminate its efforts. In the spring of 1982, the government chose to scale up the atomic separation process to a major manufacturing effort.

In the atomic separation process, the uranium is heated to 2,500° Celsius to obtain a dense atomic vapor. The vapor is irradiated with three separate red laser beams, each adjusted to the wavelengths of three stepwise transitions in the atoms of uranium-235. Irradiation causes the atoms to ionize. The positively charged ions are collected at the surface of an electrode, and the condensed, enriched material is removed from the collecting surface at regular intervals. The three red laser beams are produced by tunable dye lasers pumped by copper vapor lasers operating in the yellow/green regions of the spectrum.

In the molecular process, uranium is converted into gaseous uranium hexafluoride (UF_6) and mixed with a buffer gas. The mixture is cooled by expanding it through a supersonic nozzle to narrow the linewidths of the molecular transitions in UF_6 so that those in the two isotopes do not overlap and selective excitation can occur. Then, the mixture is irradiated with an infrared laser. The laser-excited UF_6 molecules containing uranium-235 are irradiated with a xenon chloride laser, which dissociates them into UF_5 molecules. The UF_5 precipitates at the bottom of the chamber and is periodically collected in solid form and reconverted to atomic uranium-235.

Apparently, cost played a major role in determining the selection of the atomic process for scale-up. Isotope enrichment is measured in cost per separative-work-unit, or SWU. The atomic process was projected to cost approximately \$30/SWU and the molecular process more nearly \$50/SWU. While the U.S. government has chosen the atomic process, the molecular process still has supporters. Siemens, a private company in West Germany, and URANIT, an institute of the West German government, are both still working on the molecular process.

APPLICATIONS OF LASERS

Industrial Applications

Industrial applications were among the earliest practical uses of lasers. In the very early days of ruby lasers, around 1960, output power was often measured by the number of razor blades that the beam could pierce. Pulsed lasers, such as the ruby laser, were immediately seized upon for drilling small holes in materials (diamond, for example) too hard to be worked conventionally except with the greatest difficulty. Industrial use might have remained at a relatively low level, however, except for the discovery, in 1964, of the carbon dioxide laser, which offered large amounts of continuous-wave power at high efficiency.

The ability to heat a material locally in a remotely controlled environment is one of the key attractions of using lasers for materials processing. Very high cw powers are now routinely available from carbon dioxide lasers, and moderate amounts of cw powers are available from various solid-state lasers. These devices are used in such processes as drilling, cutting, welding, and surface treatment, including hardening and annealing of metallic objects and annealing semiconductors for amorphous-to-crystalline phase transformation.

In many of these uses, the ease with which laser-processing schemes can be incorporated into computer-controlled fabrication technology makes them candidates for displacing conventional processing schemes. In many other applications, such as cutting and welding titanium, laser processing is almost the only technique that works well. Laser processing now includes applications as mundane as piercing holes in baby-bottle nipples and cigarette paper and cutting cloth in the clothing industry, and as esoteric as punching tiny holes in bismuth, tellurium, and other films for high-density, optical storage of information, as in video discs.

Carbon dioxide lasers capable of producing 1 to 10 kilowatts of cw power and neodymium-doped yttrium aluminum garnet lasers producing 100 W of cw power have become commonplace in many critical processing steps in manufacturing (see figure 7). In any case, the revolution caused by the advent of lasers in the industrial arena is likely to speed up in the near future.

Defense-Related Applications

In no other potential field of use for coherent light have more imaginative and often more far-out suggestions been generated than in defense. The concept of light beams as weapons preceded the invention of lasers by many centuries; hence, the laser was greeted enthusiastically

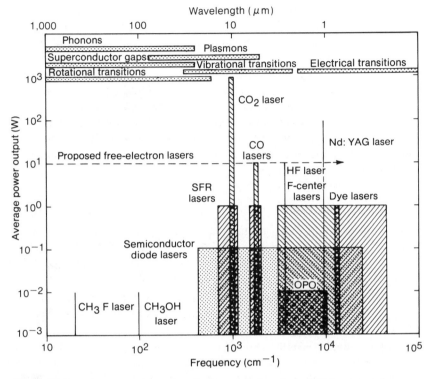

Figure 7 Power output and wavelength characteristics of some of the present lasers. In the far infrared range, there are a large number of discrete laser wavelengths at which power in the milliwatt range can be produced. [SOURCE: Bell Laboratories.]

by the defense community. Practical, laser-based weapons have yet to appear, but defense-oriented work has resulted in equally important concepts, such as laser radar and laser target designators.

Weapons applications have been pursued vigorously. Their implementation, however, is slowed by questions about the reliability and cost effectiveness of using photons to damage materials. The interaction of high-intensity laser radiation with materials, whether the skin of an intercontinental ballistic missile (ICBM) or the detector array in a reconnaissance satellite, is reasonably well understood. It is also possible to define the parameters needed to destroy a given offensive system, whether a tank, an antisatellite satellite, or an incoming ICBM. The continuing emphasis on high-energy lasers, design criteria for large optical elements, adaptive optics to combat the effects of intervening atmospheric turbulence, pointing the laser, and tracking targets is leading to potentially useful solutions. But defense against ICBM's and satellites is complicated by many factors, including the general ineffectiveness of laser weapons in the presence of dense cloud cover. Orbiting laser weapons clearly have advantages for such uses. The scheme is hampered, however, by the conflict between the choice of lasers appropriate for an overall strategy as opposed to a specific predetermined scenario, and by the complicated strategy for defending such space-based lasers against nuclear and other counterattacks. In spite of these and many other difficulties, cost effectiveness being not the least of them, the concept of laser weapons is far from moribund. Progress is likely to come in unexpected areas, involving specialized applications, and from pursuing different philosophical approaches to defense in a space-based environment.

Medical Applications

The ability to focus coherent light of sizable power within a small area has attracted the biomedical community to lasers from the very beginning of the laser era. In fact, focused light from xenon arc lamps was being used in ophthalmic surgery many years before the laser was invented. And who can forget the 1964 movie where James Bond was threatened with "radical lobotomy" with a laser years before one capable of such surgery was developed?

In general, the biomedical applications of lasers and coherent light fall into two major categories, diagnosis and surgery. The progress made in both is astounding given the conservative nature of the medical community as a whole, yet the surface has barely been scratched.

Diagnostic applications have their roots in chemistry and physics. They include such techniques as fluorescence tagging and identification of biomolecules and the counting of white blood cells in real time. Much progress has been made in clinical immunology through the use of laser cytofluorometry. This technique is used for immunofluorescence studies of tissue sections, and flow cytofluorometry is used to identify ribosenucleic acid (RNA) and deoxyribosenucleic acid (DNA) in blood plasma and abnormal cells in Pap smears, as well as to separate white cells from blood and to study cellular kinetics in leukemia. Use of a carbon dioxide laser for noninvasive analysis of blood for sugar, urea, and lipids has been demonstrated. The beam penetrates the skin only a fraction of a millimeter, and the returning radiation provides the analytical data. Laser Doppler techniques hold great promise for microscopic *in vivo* studies of blood velocity. Medical diagnosis will benefit enormously from the development of new lasers in the short-wavelength (vacuum ultraviolet) region.

The driving force behind the acceptance of the laser as a surgical tool is its ability to vaporize and remove biological tissue without physical contact. Only the cw lasers are generally considered appropriate for surgical use. Detailed analyses indicate that the smoke plume from the interaction of pulsed laser radiation with tissue often contains viable cells and viable RNA and DNA molecules and thus might spread disease to surrounding healthy tissue. The plumes from cw lasers, particularly carbon dioxide lasers, contain no viable cells, RNA, or DNA.

The cw lasers used in surgery are of two types: those relying on color-specific interaction with tissue and those relying on color-independent interaction. An example of the first is the argon ion laser in the blue region of the spectrum used to correct retinal detachment. The visible laser radiation propagates through the cornea, the lens, and the intervening vitreous humor in the ocular cavity and deposits its energy only in tissues containing the red pigment hemoglobin. Another example of color-specific laser surgery is the use of argon ion or ruby lasers on the skin to remove tattoos and birthmarks, basal cell carcinomas, pigmented epitheliomas, lesions of hemorrhagic sarcomas, melanomas, and malignant blue nevi.

For general surgery, however, the laser-tissue interaction should be color-independent. Light from a carbon dioxide laser at a wavelength of 10.6 μm is ideal because it is absorbed strongly by water, which generally constitutes about 90 percent of biological tissue. Of the three principal lasers used in surgery, the carbon dioxide laser provides the most accurate depth of incision and, therefore, is the most widely used. In addition, it has all the advantages of laser surgery: considerably less bleeding than in conventional surgery because small blood vessels are sealed immediately upon being cut; sealing of lymphatic vessels, which reduces the danger of spread of disease; no postoperative edema; and minimal danger of spread of malignancy during surgery. These and other reasons have made carbon dioxide the laser of choice in tumor surgery; head, neck, and breast surgery; neurosurgery; ear, nose, and throat surgery; gynecological surgery for such problems as cervical intraepithelial neoplasia, vaginal neoplasia, and focal carcinoma of the cervix; and plastic surgery. The carbon dioxide laser has reduced the cost of gynecological surgery substantially by permitting many procedures that previously involved hospitalization to be done during an office visit.

Surgery with the carbon dioxide laser is a prime example of a new modality that was accepted with open arms. But, problems remain. A major one is the lack of optical waveguides suitable for transmitting the beam from the laser to the patient for precision surgery. In the absence of such waveguides, articulated arms with corner-mirror reflectors are commonly used. These devices suffer, however, from inaccuracies in pointing. Recently, accuracy has been improved substantially with the invention of an articulated arm using hollow quartz tubes as waveguides. This is likely to revolutionize surgery with carbon dioxide lasers in the next five years.

Outlook

Lasers have progressed remarkably during the past quarter century and show no sign of slowing. The devices are used today in basic science, communications, industry, defense, and medicine, and new applications in all of these areas can be anticipated.

Existing lasers meet the needs of science and technology over much of the electromagnetic spectrum. New types are needed, however, especially types that produce light of shorter wavelengths and pulse durations than are available today. Tunable lasers should be able to cover the low end of the spectrum to wavelengths as short as 50 nm within the next few years, and pulse durations shorter than 30 fs are in prospect.

Existing lasers are steadily expanding our knowledge of matter, and new types expected soon will provide even deeper understanding. High-resolution laser spectroscopy, for example, is helping to confirm the ideas of quantum electrodynamics. The same methods are contributing steadily to our knowledge of the structure and behavior of solid materials used in electronic and other kinds of devices.

The technology of lightwave communications—the transmission of information by laser light through small optical fibers—is advancing rapidly. Many systems using first-generation technology are operating, and second-generation technology is well along in development. A prototype system has operated over 101 km of optical fiber at a high data rate and essentially zero error rate.

Lasers also can be used to separate the isotopes of elements, and a process for enriching the fissionable isotope of uranium for nuclear fuel and weapons is scheduled for major scale-up. It promises to achieve in a single step the enrichment now achieved in the costly, multiple-step, gaseous diffusion process.

Lasers are used in industry for many purposes, including drilling, cutting, welding, and surface treatment. In many of these uses, laser processing schemes are peculiarly well adapted to computer-controlled technology. The revolution caused by the advent of lasers in industry is likely to speed up in the near future.

Defense applications of lasers include laser radar and laser target designators. The use of lasers as weapons is complicated by many difficulties, but the field is extremely active. Progress is likely to come in unexpected areas involving specialized applications.

Lasers have made considerable progress in medicine, for both diagnosis and surgery, yet the surface has barely been scratched. In addition to many existing applications, the use of a carbon dioxide laser for noninvasive analysis of blood for sugar, urea, and lipids has been demonstrated. The invention of an articulated arm that uses hollow quartz tubes as waveguides is likely to revolutionize carbon dioxide laser surgery in the next five years.

BIBLIOGRAPHY

H. C. Casey, Jr. and M. B. Panish. *Heterostructure Lasers*. New York: Academic Press, Inc., 1978.

S. Chu and A. P. Mills, Jr. "Excitation of the Positronium $1^3S_1 \rightarrow 2^3S_1$ Two-Photon Transition," *Physical Review Letters*, Vol. 48, No. 19 (May 10, 1982), pp. 1333–1336.

W. W. Duley. *CO_2 Lasers—Effects and Applications*. New York: Academic Press, Inc., 1978.

R. L. Fork et al. *The Proceedings of the Conference on Picosecond Phenomena, III*. New York: Springer-Verlag, to be published.

Laser Surgery. Proceedings of the 3rd International Congress for Laser Surgery. Edited by I. Kaplan and P. W. Ascher. Graz, Austria: 1979.

Lasers in Industry. Edited by F. F. Charshan. New York: Van Nostrand Reinhold Co., 1972.

T. J. McIlrath and R. R. Freeman. *Conference Proceedings No. 90 of Topical Conference on Laser Techniques for Extreme Ultraviolet Spectroscopy. Optical Sciences and Engineering*. Volume 2. New York: American Institute of Physics, 1982.

C. K. N. Patel. "High-Power Carbon Dioxide Lasers," *Scientific American*, Vol. 219, No. 2 (August 1968), pp. 22–33.

C. K. N. Patel and E. D. Shaw. "Free-Electron Laser—Its Use in Condensed Matter Physics Research," in *Novel Materials and Techniques in Condensed Matter*. Edited by G. W. Crabtree and P. Vashishta. New York: Elsevier Science Publishing Co. Inc., 1982, pp. 179–192.

8

The Next Generation of Robots

Robotics is often counted as one of the branches of artificial intelligence. As such, it partakes of the exalted aspirations of that millennial field. However, its immediate prospects are bound up with far more prosaic matters, notably with the rapidly growing industrial efforts to improve productivity by extending and "roboticizing" the present techniques of automation.

For the purposes of this chapter, robots can be defined as computer-controlled devices that reproduce human sensory, manipulative, and self-transport abilities well enough to perform useful work. (Of course, robot sensors sometimes surpass human capabilities in certain applications, and they may work in very different ways.) Proceeding by anthropomorphic analogy, the components of these machines can be grouped into six main categories:

- Arms—robot manipulators.
- Legs—robot vehicles.
- Eyes—robot vision systems.
- Ears—computerized speech recognition systems.
- Touch—tactile sensors and artificial "skins."
- Smell—smoke detectors and chemical sensors of other kinds.

There are also other, less human kinds of sensors. Commercial ro-

◀ A time-lapse photograph of a robot manipulator assembling gears. The device can sense and correct improper seating of the gears. [IBM Systems Products Division.]

bot systems are available in all of these categories, and systems that integrate multiple capabilities are beginning to appear on the market.

In addition, there are many kinds of nonrobot systems that handle tasks requiring some degree of intelligence, for example, "expert systems" of various kinds. These systems were discussed in the second *Outlook for Science and Technology: The Next Five Years* (W. H. Freeman and Company, 1981, pp. 747–753).

HARDWARE AND SOFTWARE LIMITATIONS OF CURRENT ROBOT SYSTEMS

Both the sensory capabilities of robots and their ability to deal with unexpected events are as yet quite limited. For this reason, robots presently are effective only in highly structured environments in which the position and path of motion of all objects are known fairly precisely at all times. This largely confines robots to industrial environments, and mostly to work with hard materials (metal, wood, hard plastics), rather than with soft, flexible objects (cloth, vinyl) that are hard to control.

In spite of these limitations, robotics is expanding. This is partly the result of its mastery of such commercially significant activities as spot welding and spray painting and an accumulation of technical expertise over the past decade. It is also a consequence of the microelectronics revolution which already has reduced the cost of computer control drastically, and which, in years to come, should produce many integrated "miracle chips," embodying advanced sensory and end-effector control functions. The combination of electronics, technology, and software science represented by robotics can be expected to lead industry into an age of electronic factories with drastically diminished manufacturing labor forces. However, all of this lies some decades in the future. Even the use of robots to perform the majority of industrial operations is decades away. For the near term, the potential applications of robots are extensive, but not unlimited.

APPLICATIONS: ECONOMIC CONSIDERATIONS

The automated industrial environment is one in which workpieces, that is, items being manufactured, move along controlled paths through a variety of steps, such as crimping, shaping, welding, melting, cutting, stamping, spraying, painting, and assembling. As workpieces move along an automated manufacturing line, they are placed accurately on pallets or pushed into positions between fixed walls, pipes, and chutes, as well as rods, plates, cams, and other moving members. This requires detailed sculpturing of the geometric environment in which the workpieces move. The specialized fittings and mechanical tooling demanded by this process make fixed factory automation very expensive and ill-equipped to cope with changing product lines. Robotic technology aims to use a few standardized but adaptive mechanisms, of which the robot manipulator is the prototype, to replace these expensive special purpose fittings. Of course, robot manipulators will continue to use many of the tools presently employed in automated manufacture.

Robot Manipulators

Generally speaking, robot manipulators (for example, industrial screwdrivers) perform with high efficiency special motions that grasp or otherwise acquire objects (such as electromagnets and vacuum lifts) or that apply special physical processes to workpieces (for example, spray gun, welding gun, industrial shears, stamping press, drill press).

Fixed automation can be expected to remain more advantageous for high-volume manufacture than programmable robot equipment. This is because specially designed mechanical equipment can move more rapidly than general-purpose robot manipulators, and it is therefore more productive once the high initial cost of setting up an automated production line has been amortized. On the other hand, for manufacturing individual items or very small batches, manual production often will be less expensive than a robot control program. Thus, the area most favorable to robotics is medium-quantity batch manufacture; in other words, the manufacture of items in batches numbering several hundred to several thousand. In this connection, it is quite important that robot systems be flexible and easy to adapt to new applications. The more the setup costs associated with typical applications can be reduced, the smaller will be the size of the least robot manufacturing run that remains economical.

Robot techniques also will be advantageous when larger numbers of similar but not identical items need to be manufactured, provided that the variations between items are not too large to be accommodated comfortably without very sophisticated computer control. The furniture manufacturing industry illustrates this situation: many items of furniture, especially fine furniture, are produced in relatively small batches as orders for specific designs are received and as shipments of wood with matching grains come in.

Today's robot manipulators are usually sedentary, stand-alone arms with limited precision and very little sensory capability. Many improvements in these relatively primitive devices are possible but, even in their present state, they support applications with growing economic values. The most important current industrial applications of robot equipment are handling materials, loading/unloading machines, trans-

Table 1 Breakdown of shipments of
playback robots by work process,
March–September, 1979

Work process	Number of units	Percentage of value
Spot welding	57.1%	45.1%
Arc welding	18.8	26.0
Spray painting	11.3	17.8
Other	12.8	11.1

SOURCE: Paul Aron. "Robotics in Japan," *Paul Aron Reports*, No. 22 (July 3, 1980). Published by Daiwa Securities America, Inc., New York.

port (surrogate conveyor), palletizing/depalletizing/kit packing, processing and fabrication, welding (spot and arc), spray painting, drilling, assembly (parts mating), and testing (dimensional, continuity).

Table 1 describes the ways in which one of the simpler classes of robot devices is used. This class consists of the "playback robots," which simply repeat a sequence of motions through which they have been led. (These simple robots can be seen in major automobile plants.)

ROBOTS IN THE UNITED STATES AND JAPAN

Although France, West Germany, and Sweden all have active robotic research groups and are introducing robots rapidly into industrial settings, robot use is most advanced in Japan. It is sobering to reflect that, even though much more than half of all robotics research and development before 1975 was performed in the United States, Japan now leads the United States in applications. Among other things, this illustrates the fundamental importance of technology transfer mechanisms deeply rooted in a nation's educational and economic systems. Table 2 shows how far behind Japan the United States has fallen in the wider user of industrial robots.

Most (nearly 70 percent) of the installed Japanese robot manipulators are of the very rudimentary "fixed sequence" class; in other words, they execute repetitively a sequence of motions that is not easily changed. "Playback" robots differ from these in that they can be "taught" new motions by being led manually through the sequence of points to be traversed. These constitute roughly 8 percent of installed Japanese robots, but 16 percent (by dollar value) of the Japanese robots shipped in 1979. More sophisticated robots of various kinds constitute roughly 12 percent of those shipped in 1979.

As background to these figures, it should be noted that manufacturing accounts for roughly 21 percent, and durable goods manufacture for roughly 13 percent, of the U.S. gross national product.

Table 2 Comparison of industrial robots in the U.S. and Japan, 1980 (using U.S. definition*)

Robots	Japan	U.S.
Production in units	3,200	1,269
Production in value ($ million)	180	100
Installed operating units	11,250	4,370

*Japan recognizes six classes of robots, while the U.S. recognizes four. The U.S. definition covers variable-sequence robots, playback robots, numerically controlled robots, and independent robots. Japan adds manual-manipulator and fixed-sequence robots.

SOURCE: Paul Aron. "Robots Revisited: One Year Later," *Paul Aron Reports*, No. 25 (July 21, 1981). Published by Daiwa Securities America, Inc., New York.

LIMITS ON TODAY'S ROBOTS

As already noted, currently available robots are quite limited both in their sensory capabilities and in their ability to deal with unforeseen contingencies. A corollary is that, as items to be processed enter the robot workspace, either they must be in their proper order as a result of preceding operations, or order must be imposed by passing them through parts feeders or chutes of an appropriate geometry. Thus, as these workpieces move in a roboticized factory, either control over their position must be maintained, or reorientation operations must be performed repeatedly. A central aim of industrial robotic research is to relax these constraints—that is, to find ways of dealing with less and less structured environments. Accomplishing this will require development of better sensors and improvement of procedures for analyzing sensor-generated data, object recognition algorithms, environment-modeling software, and computerized replanning techniques. The "bin picking" problem, the problem of locating a part in a disordered bin of parts so that it can be lifted out and given a prespecified orientation, is typical of the problems that designers face in dealing with disorder.

Characteristics of Manipulators

Current, general-purpose robot manipulators have one arm with three to six independently controllable "joints" (involving motor, encoder, and controller) plus a gripper that can be opened or closed. Dozens of manufacturers are offering manipulators of this kind. Generally, these arms are not mobile, but a few experimental mobile robots have been built. However, today's robot carts are generally armless, stand-alone vehicles that follow preset paths. Typical uses are delivering office mail or hospital meals and linen.

The six degrees of freedom that are typical of current robot manipulators suffice to position the robot's gripper, or a tool or workpiece that it is holding, at an arbitrary point within the robot's accessible space, with an arbitrary rotational orientation. The gripping hand mounted at the end of a manipulator arm is often equipped with a few simple sensors, for example, strain gauges that sense contact with external bodies and grip forces, or photocells that detect the presence of an object between the gripper fingers. In some setups, one or more television cameras are mounted in positions that allow them to observe the robot arm or hand and the objects that the manipulator approaches; information derived from analysis of the images is passed to the program controlling the robot. Control is ordinarily exercised by an inexpensive mini- or microcomputer.

The "reach" of a typical robot manipulator may vary from 1 to as many as 10 feet; the "payload" that it can lift ranges from a few ounces to several thousand pounds.

Figure 1 shows such a robot manipulator. Manipulators vary in price from a few thousand dollars for small, simple, slow-moving arms without sensory capabilities, intended for light educational use, to roughly a hundred thousand dollars for fast, sensor-equipped industrial systems capable of highly precise mechanical movements (for example, returning to within 0.01 of an inch of a prespecified position). The working volume of a large industrial manipulator, in other words, the volume of space it can reach, may be as fast as a cube 10 feet on a side. Speeds as high as 5 feet per second are not unusual.

Although the sophistication of robot control systems can be expected to increase rapidly because of the growing availability of powerful microprocessors, much of the industrial robot equipment in use today cannot respond flexibly to changing external situations. Manipulators of the "playback" type, which simply store and repeat some specified sequence of motions, represent the extreme. More flexible systems include a mini- or microcomputer for control. In such systems, control programs written in a reasonably powerful robot programming language are stored in the computer's memory and determine the manipulator's sequence of motions. Special statements in the robot language allow information to be read from sensors and responses conditional on this sensed information to be programmed. However, even in using a programmable robot, it is often convenient to describe both the main outlines of the sequence of motions that it is to perform and the key points that these motions are to traverse, by moving the arm through these desired positions manually. For this reason, robot manipulators are generally equipped with a "teach box" with keys that allow the arm to be moved or rotated by hand. The robot's control computer then acquires the manually guided motions and integrates them with a more comprehensive program that also involves the use of sensors and sophisticated conditional responses.

Figure 1 Puma 600 manipulator. [Unimation Inc.]

As long as a manipulator arm can be moved through free space up to the precise point at which it will make contact with an external body, determination of the manner in which it is to move is relatively straightforward. Although challenging problems of dynamic control do arise when the manipulator is to be moved rapidly, especially if it is simultaneously grasping an object of appreciable mass or if it must operate in a moving coordinate frame, generally, control of unimpeded motions is not difficult. However, it is impossible to maintain absolutely precise information concerning the position of a manipulator and its workpieces at all times, if only because manipulator arms themselves will deform slightly as they move, or become worn or slightly ill-adjusted. Therefore, one needs to deal with the phenomena that arise when an object firmly gripped by a manipulator comes into contact with an object fixed in some other coordinate frame. Here, purely geometric control becomes infeasible since, if it touches an object, a manipulator moving along a rigidly prescribed path will break either the object or itself. Consider the problem of inserting a peg into a hole; if the motion of the peg is perfectly independent of the forces exerted on it by the walls of the hole, any geometric imprecision will jam the peg either at the mouth of the hole or against one of the walls.

Thus, as bodies come into contact, robotics leaves the purely geometric domain and must deal with problems of force-sensing and with hybrid motions guided, in part, geometrically, but also by compliance with external forces. An important step toward more flexible robot systems would be for the control software furnished with manipulators to provide "force-controlled" motion commands; using these, one could, for example, easily cause a manipulator to paint a stripe on the exterior surface of an automobile or apply adhesive to an aircraft window frame. However, although techniques for doing this have been investigated theoretically and demonstrated in research laboratories, motion control at this level of sophistication is not yet available as a standard feature of commercial robot equipment.

ROBOT PROGRAMMING LANGUAGES

If robots are to be applied widely in industry, they must be easy to use. Potential users forced to deal with rigid, hard-to-understand robot control languages will be discouraged from applying the new technology. For this reason, more powerful and user-friendly robot languages are essential.

Although we are far from knowing how to describe those built-in operations that would be most desirable for such a language, the following discussion of significant operations should help to summarize the present state of these languages and to anticipate some of the things that future robot languages are likely to provide.

Available Languages

The relatively undeveloped state of robotics is illustrated vividly by the fact that current robotics languages support only a handful of the primitive operations that are desirable. (In this context, "primitive" refers to built-in, efficiently implemented operations which can be initiated by a single statement of the language.) The primitive facilities that now exist in at least one commercially available robot language are roughly the following:

1. Manipulator motions can be described and controlled in XYZ axes fixed in space, or in frames defined relative to an object grasped by the manipulator. Motions passing through a known sequence of positions/orientations, with known speed, can be specified, typically by commands having the form:

MOVE ARM THRU POINTS P_1, . . . , P_n AT SPEED S.

2. Motions can be controlled and manipulator positions determined quite precisely. The manipulator can be put into a manually guided mode and moved to any desired position. This position can then be measured by the controlling computer, stored, and the manipulator

returned to automatic mode. (As already noted, this "teach" mode or "guide-through-the-motions" approach is, in fact, the easiest and most commonly used mode of programming ordinary robot applications.)

3. Several geometric and symbolic computations (for example, transportation of geometric data from one Cartesian frame to another, and generation of code from manually "taught" motions) are supported by the more advanced commercial robotics languages. Using these facilities, one can, for example, teach a robot the position of various points on an automobile body by guiding the manipulator manually to these points, and then easily cause the robot to move to these same body points, even if the body, moving down an assembly line, is presented to the manipulator at a different angle. A typical command reflecting this general capability might have as its form:

MOVE ARM 3 INCHES FORWARD IN FRAME OF TOOL.

4. Tactile and visually sensed events can be detected. These can be used either to trigger interrupt routines or to terminate motions. User-defined events generated by nonstandard sensors are provided for also; for example, a signal from one or more photocells mounted near a manipulator can halt or redirect the manipulator's activity immediately (as safety concerns might require). A typical command used for this purpose might read:

WHENEVER SWITCH-1-CLOSED IS DETECTED, EXECUTE SAFETY PROCEDURE.

5. Parallel processes (capable of controlling the activity of multiple arms or other effectuators) can be defined, activated, suspended, synchronized, etc.

6. Some image-analysis software is available. This is beginning to be integrated with manipulator control software, facilitating the construction of combined hand–eye systems.

The last item brings us to the forefront of robot technology now available commercially.

Desirable Extensions to Robot Programming Languages

The addition of new statements to robot programming languages can be expected to reflect advances in robotic technology rather faithfully. The following are some of the additional control language facilities that would be most desirable:

1. Statements that cause a manipulator arm to move between specified positions while automatically avoiding certain obstacles or unde-

sirable arm configurations. For arms with more than the standard six degrees of freedom, this may involve the capacity to "reach around" objects in a manner impossible for a standard arm. A robot control language designed for these purposes also should be capable of dealing correctly with situations in which the arm is gripping some object that will move with the arm. For example, one would like to be able to use a command like:

MOVE TOOL FROM FRONT-OF-BOX TO BACK-OF-BOX.

leaving it to the robot to figure out the path necessary to avoid bumping into any obstacle.

2. Motion control statements specifying specialized "wobbling" or "twisting" motions useful for overcoming friction. For example:

PUSH PEG-1 1 INCH FORWARD WHILE TWISTING.

3. Statements for managing force-controlled motions during which some of a manipulator's geometric coordinates move along specified paths at specified rates while other degrees of freedom adjust to the environment via sensed forces. Such statements might specify motion in contact with surfaces, edges, seams, and so on. An example would be:

SLIDE FINGER ALONG SEAM MAINTAINING 1 POUND VERTICAL FORCE.

4. Statements for guiding a self-locomoting robot.

5. Statements assisting in grip management, i.e., in determining the minimum gripping force that must be exerted in order to prevent a body of a given weight, gripped at known points, from slipping. Also necessary is some way of automatically causing the grip on an object to tighten when the object begins to slip.

6. Statements for managing a deformable or multifingered hand that is to grasp an object "geometrically" rather than by friction, enabling it to "surround" the body so that it cannot fall or otherwise escape without passing through part of the hand. An example might be:

GRASP PASSING MIDDLE FINGER UNDER ROD.

7. Statements for coordinating the frames of general objects moving relative to each other, including objects whose position can be sensed but not affected by a robot's control computer. These would include statements for simultaneously monitoring the state of many sensors, for coordinating the activity of multiple robots, and for assuring that, when multiple robots enter each other's immediate vicinity, collisions are avoided. An example might be:

MOVE HAND-1 TOWARD FRONT-EDGE-OF-BALL,
APPROACHING AT RATE 1 FOOT-PER-SECOND

when the "ball" in question is being held by another robot hand.

8. Statements for dynamic control of a manipulator, e.g., for causing a manipulator to strike a calibrated blow.

9. Statements for managing the activity of a small robot attached to or held by a large robot.

10. Facilities for modeling relations of attachment and of gravitational support.

ROBOT SENSORS

A variety of attached sensor devices enables robots to respond to their environment. These include ultrasonic range sensors, tactile sensors of various kinds, and visual sensors. The following discussion will concentrate on visual sensors, which are particularly important because of the high speed and nonintrusive nature of vision. However, tactile sensing can be more important than vision in some applications, for example, in the assembly of tightly fitting mechanical parts. The elastic and frictional forces critical in such applications vary so rapidly with changes in relative part position that little understanding of what is happening when a part jams during assembly can be gained by visual inspection. Instead, one needs to use strain gauges or other tactile elements to measure and respond directly to the forces that develop during assembly. This explains the importance of such devices as the Draper Laboratory's Remote Center Compliance (described below) and the strain-gauge instrumented version of it now being developed.

Structure and Characteristics of a Vision System

Image acquisition is, of course, a necessary prerequisite to all else. Two approaches are used currently. These are uniform illumination (like that most comfortable for the human viewer) and "structured light." At present, uniform-illumination schemes, which exploit readily available television monitors to form digitized scene images, are the most common. The capabilities of such a vision system will, of course, be constrained by the quality of the images with which it works and the rate at which basic image-processing operations can be applied. The spatial resolution of the solid-state video sensors used in uniform-illumination robot vision systems of this kind is increasing rapidly. Inexpensive devices for acquiring two-dimensional arrays of 256×256 pixels (elementary "dots" in a picture) in less than one millisecond, and arrays of 512 × 512 pixels or finer at video rates, are available commercially, as are somewhat more expensive devices for acquiring images with up to 2,000 × 1,000 pixels.

Because of the human inability to detect video display flicker above 60 Hertz, most video systems are designed to process a single image in about 25 milliseconds. High-performance "video stream processors" are designed to perform single and binary frame operations at that same rate.

Structured-light vision schemes illuminate a scene by one or more narrow planes of light, each of which lights up some narrow, broken curve on the bodies that make up the scene (see figure 2). Since the plane of illumination is controlled, and since a line of sight drawn in a known direction from the viewing camera will intersect this plane at just one point, the position in three-dimensional space of each point on the curve can be calculated.

By sweeping the illuminated plane at a controlled rate over a scene, one can determine the full three-dimensional location of each body surface in the scene. Even though measurement imprecisions, unwanted specular reflections, and problems of camera calibration all complicate the mathematical simplicity of this scheme, the structured-light approach, which is currently used in a few industrial locations and is being refined by researchers at Stanford Research Institute, the National Bureau of Standards, and elsewhere, seems promising.

Once an image (either a two-dimensional image or the kind of three-dimensional image of a surface that a structured-light scheme is capable of producing) has been acquired, one needs to use it to locate and recognize the bodies present in the acquired scene. The complexity of this "scene analysis" depends on the assumptions one can make concerning the scene to be analyzed. If the scene is known to contain only one object and if the object is being viewed from one of a known finite number of positions, relatively simple processing will locate and identify the body. The technique typically used to handle simple cases is to divide the digitized picture into separate "blobs" (each defined as a connected region of some characteristic threshold intensity) and to extract global features of these blobs, for example, their area, centroid, principal axes, and number and size of contained "holes." In the simplest cases, this information, which can be extracted using relatively easy bit-parallel arithmetic and Boolean operations, suffices to distinguish bodies and determine the angle from which they are being viewed.

In less simple cases, and especially if some of the bodies being viewed are partly obscured by other bodies in the same scene, more difficult analyses, which research has not yet fully reduced to practice,

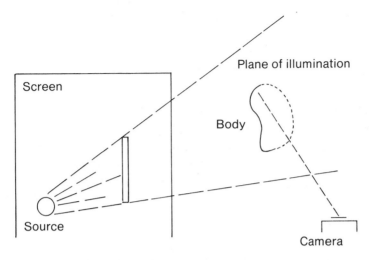

Figure 2 Structured-light approach to imaging, showing intersection of plane of illumination with line of observation. By sweeping the illuminated plane at a controlled rate over a scene, one can determine the full three-dimensional location of each body surface in the scene.

must be attempted. Basically, one tries to extract local body features (such as a person's nose seen in profile) by which an object can be identified even when most of it is not visible. A typical first step is to find such characteristic features as "edges" and "corners"; edges are defined as curves along which intensity changes rapidly, and corners as points at which the direction of an edge changes rapidly or at which several edges meet. Unambiguous detection of these image features in the presence of digital noise, breaks in perceived edges, and shadows and random bright spots is not easy, but some success has been attained by applying appropriate differencing and other picture-enhancement operations across the individual pixels of an image.

Although some of these methods have had partial success, it must be admitted that current computer vision software is of limited reliability and quite expensive computationally. Much faster and more stable picture-processing algorithms and devices are needed. Techniques for the identification of partially obscured bodies are particularly important. In order to deal with moving bodies (for example, to sense their rate of motion), higher picture-processing speeds are needed. Vision systems that can be reprogrammed easily for a wide range of applications are also desirable but, at present, it is not clear how they can be created. It is hoped that understanding can be reached concerning the algorithms most appropriate for detecting image features and how to use them to identify objects present in a scene. This will encourage the development of dense, complex, special purpose integrated circuit (VLSI) chips, which could then perform the required optical sensing operations and geometric computations very rapidly.

Applications of Computerized Image Analysis

Current Applications   Although the capabilities of robot vision systems still fall far short of the subtle feats of location and identification that the human eye performs so effortlessly, these systems now suffice for a variety of important uses. A typical application is the identification of well-spaced parts moving along a conveyor belt; after being located, the parts can be grasped and placed in a standard orientation. Inspection of manufactured parts is another application growing in economic importance: here, the two-dimensional image of an object of moderate complexity, for example, a metal casting, is examined to make sure that all of its expected subfeatures, and no others, are present. In another recent application, visual sensing is used to find seams to be traversed by a robot arc welder; three-dimensional (structured-light) vision might be particularly appropriate in this application.

Future Applications   The object-inspection techniques mentioned in the preceding paragraph have been applied to inspection of faults in printed circuit boards, integrated circuit masks, and integrated circuits themselves. However, today's robot vision systems are too slow to be cost effective in dealing with images of this complexity. Falling computation costs and the development of special chips to perform basic image-processing operations at extreme speeds can be expected to remove this limitation and thus to expand the range of objects that can be inspected economically by robot equipment.

Still more sophisticated image-processing will be required before robots can move amid the clutter of the ordinary industrial environment. Initial experiments, such as those conducted at the Stanford Artificial Intelligence Laboratory, make the problem of unraveling views of ordinary, uniformly illuminated scenes appear quite challenging. Here, however, we can hope that three-dimensional vision methods will be superior; at any rate, the range of information generated by schemes of this kind should make it easier to locate and avoid obstacles. Deeper scene analysis methods will be required for still more sophisticated applications of vision; for example, locating and grasping objects partly hidden in the clutter of a standard industrial tote bin, or identifying trash cans standing near bushes and behind trees on a suburban lawn.

OTHER RESEARCH PROBLEMS

Although robotics research seems certain to touch upon a particularly broad range of technologies and scientific disciplines, some understanding of the areas likely to be significant over the next decades can be gained by surveying the near-term requirements of industrial robotics. These include:

1. *Faster robots and increased robot accelerations.* The productivity of robot equipment depends upon the speed with which it moves. For this reason, the seemingly pedestrian problem of increasing the rates of motion and acceleration of robot manipulators is of particular economic importance. This is partly a problem in mechanical engineering. It is also a matter of combining a more sophisticated understanding of the elastic reactions of rapidly moving manipulator members with the computer control techniques to compensate for the mechanical and sensory inaccuracies of high speeds and accelerations and to suppress unwanted oscillations. The solution will require improved force sensing and more sophisticated servocodes.

2. *Improved optical, tactile, and other robot sensors.* Tactile sensing plays a particularly important role in dexterous manual assembly. The subtlety of the human tactile sense is far from being matched by the relatively crude sensors currently available with robot manipulators. It seems clear that greatly improved sensors will be required if complex assemblies, especially of fragile and deformable parts, are to be attempted, if robots are to work extensively with soft or plastic materials such as wires, cloth, or vinyl, and if more sophisticated methods of grasping are to be developed. Important work under way on tactile sensors should yield considerably improved sensors within a few years. These sensors will be able to detect not merely that a robot's finger is touching something but, also, the precise parts of the finger at which contact occurs; in addition, they will be capable of delivering at least a crude "tactile image" of the surface touched. (To see how important this is, note carefully what you do when turning a page in a book, especially with your eyes closed. A delicate tactile sensation of the manner in which the edge of the page being turned rests against the operative finger is necessary in order to slip the finger under the edge of the page at just the right time to accomplish turning.)

It is also important to develop improved proximity sensors capable of giving advance warning of impending collisions. Without adequate proximity sensors, one will never be willing to set robot arms into rapid motion in an environment that is to any degree unpredictable. It is plain that the development of better proximity sensors can contribute substantially to the productivity of robot systems and also to their flexibility.

Better tactile and other sensors will, in turn, call for more sophisticated software to manage them, a consideration that emphasizes the significance of improved programming techniques and higher performance computers to the general progress of robotics.

3. *Force-controlled motion primitives.* The motion control primitives supplied with today's robot systems are purely geometric in character, but cannot, as they stand, be used to cause a robot arm to move smoothly while it maintains contact with a curved surface of unknown shape. This ability is essential for the smooth and logically flexible adaptation of a manipulator to an environment in which the whole geometry is not known in precise detail.

Force-controlled motions are essential in manual assembly; blind workers can function in many factory roles where a worker with anesthetized hands might be useless. The demonstrated advantages of devices like the Draper Laboratory's Remote Center Compliance, referred to earlier, illustrate this. As anyone who has struggled to close a tightly fitting desk drawer knows, a long object being pushed into a closely fitting cavity can jam easily. Using geometric analysis, it is easy to see that the tendency to jam is greatly reduced if the object is pulled into the cavity rather than pushed. Of course, this is normally impossible: one could hardly crawl into a desk to pull a drawer in from its reverse side. However, the simple but ingenious arrangement of hinges used in the remote center compliance results in exerting exactly the pattern of forces that normally would have to be exerted by pulling the body from its other side. The wide industrial interest in this device points clearly to the importance of proper force control for robot manipulations. Research and development efforts that build on the success of the remote center compliance and its underlying theory are therefore likely. Ultimately, they should make force-controlled motion primitives available in the commonly used robot programming languages.

4. *Geometric modeling.* Geometric modeling refers to the use of a computer to build models of three-dimensional bodies and the constraints on their relationships. These then can be combined into a comprehensive environment model within which various kinds of kinematic and physical analyses become possible. These "computer-aided design" models are used fairly widely by designers to produce graphic images of an object being built, to generate engineering prints, and, in particularly advanced systems, to build control programs for the numerically controlled machine tools which produce the object that a designer has conceived.

The demands of robotic applications can be expected to force systems of this type to considerably higher levels of sophistication. For example, graphic systems may play an essential role in debugging robot control programs. If multiple arms and complex synchronized motions are involved, attempting to debug control programs by actually setting expensive machinery into motion may simply be too dangerous because of the possibility of collision. It would certainly be preferable to construct a purely graphic world in which one could view the motions that a control program would trigger. However, generating computer motion pictures of complex curved bodies in the volume required for this type of debugging will require geometric procedures of an efficiency and sophistication considerably beyond what is possible today.

Modeling the motion of bodies in contact, as influenced by their geometry, elasticity, and mutual friction, is a related but even more challenging problem. Ultrafast computers may be required.

5. *Robot locomotion.* Robots with legs (or wheels, or half-tracks), which are able to move through their working environment, are desirable for a variety of applications. One interesting case might be a "brachiating" robot that clambers, monkey-fashion, over the surface of a spacecraft (or submarine vessel) to make repairs, or over a space station it is building.

Current applications of self-moving robots are largely rudimentary, and the few interesting robots of this type that have been built exist only in research laboratories, where they have been used to investigate some of the many problems connected with robot locomotion. One of these is avoiding obstacles and finding paths in an environment not known ahead of time; aside from its planning aspects, this is basically a problem in visual analysis and range-sensing.

The problem of legged locomotion has been studied by groups in the Soviet Union and the United States. The most active U.S. group is that at Ohio State University, where researchers have constructed a moving "hexapod," over which experimental control is exercised by a simple "joystick" used to elicit forward, sideways, or turning locomotion (see figure 3). As of 1982, the Ohio State work has reached the point at which the hexapod can stride successfully over a flat floor at a speed of 20 feet per minute; work on sensory adaptation to an irregular terrain is beginning.

A group at Carnegie-Mellon University has begun studying the more complex problem of dynamic stabilization of statically unstable walking robots. As these basic kinematic/dynamic studies achieve success, robot "leg" systems will become available. The "legs" will be controlled relatively simply, for example, by geometrically specifying some desired path over terrain, and also by defining the manner in

Figure 3   Ohio State "hexapod." [Communications Services, The Ohio State University.]

which the robot is to adapt to terrain irregularities, the way in which proprioceptive "balance" signals indicating incipient falls are to be handled, and so on. Like all other primitive robot capabilities, this should result in the addition of appropriate new statements to the languages used to program higher level robot activities.

The program controlling a self-moving robot will have to keep track of the robot's constantly changing position and use this information to handle references to objects fixed in the robot's environment. Although known geometric techniques can be used, no commercially available robot programming system offers this feature.

6. *Improved robot programming techniques.* A considerable body of literature on industrial assembly gives detailed directions for producing a great many common manufactured items. Some way of automatically translating these manuals into robot assembly programs would be ideal but, unfortunately, this objective far exceeds the capability of today's robotics programming languages. For anything close to the language of standard industrial assembly manuals to be accepted as robot control input, much more sophisticated languages will be required. The compilers of such languages will have to incorporate knowledge of the part/subpart structure of partially assembled objects, as well as routines capable of planning the way in which objects can be grasped, moved without collision through a cluttered environment, and inserted into a constrained position within a larger assembly.

This level of programming sophistication only becomes feasible if a robot system can maintain either a detailed model of the environment with which it is dealing, and keep this model up to date through a complex sequence of manipulations, or if the robot can acquire and refresh such a model through visual and tactile analyses of its environment. Although no robotic language with this degree of sophistication has been produced, such languages have been projected, for example, in the work on AUTOPASS at IBM, and its Stanford University, Massachusetts Institute of Technology (MIT), and University of Edinburgh relatives, AL, LAMA, and RAPT.

It should be noted that the implementation of such sophisticated languages will require the solution of many complex mathematical and geometric problems. A basic one is that of planning collision-free motions of three-dimensional bodies through obstacle-filled environments. This problem, studied by researchers at the California Institute of Technology, IBM, MIT, New York University, and elsewhere, has been brought to a preliminary stage of solution but, from the practical point of view, the work merely reveals the complexity of the computations that motion planning involves and the importance of seeking much more efficient motion-planning schemes.

Work reported by MIT and other laboratories also suggests possibilities for more advanced software that plans robot activity. Working within a simulated world of blocks, MIT researchers constructed a program that could combine geometric knowledge of the collection of blocks given it with an understanding of the desired final assembly to produce a fully sequenced assembly plan. This demonstration program was even capable of using some of the blocks available to it to construct fixtures useful in the assembly of the remaining blocks.

7. *In-depth studies of important current applications.* Spot welding by robots has become routine, and attention is turning now to the more complex physical problems associated with continuous arc welding, where proper control of welder robots requires some understanding of the thermodynamics of the liquid–solid arc pool. Ways of developing specialized robots to work in environments hazardous or otherwise inaccessible to humans, such as high-purity clean rooms, deep-sea environments, nuclear reactors, and space, also require detailed study and will sometimes raise very complex dynamic and other problems. Improving robot tactile sensors and motion control software to the point at which robots could work successfully with soft industrial materials, for example, leather, foam rubber and/or vinyl furniture coverings, is of obvious economic importance to such industries as furniture and shoe manufacture.

FUTURE APPLICATIONS

The Household Robot

Even the relatively simple problems of industrial robotics are quite challenging, and we are far from being able to create useful, general-purpose household robots. Nevertheless, this intriguing and often-discussed possibility is worth examining—not because such applications are urgent, but simply to form some impression of the technical problems that would have to be faced in order for household robots to become feasible. In a sufficiently advanced technology, one might hope to apply robots to ordinary household tasks: cleaning, returning items to their storage positions, unpacking groceries, vacuuming, dusting, folding laundry, making beds, and so on.

Major technological advances will be required before robots can perform a spectrum of familiar tasks. The household environment is much more varied than the industrial environment, and it is not nearly as controllable. Rather than being able to deal with a very limited number of workpieces whose geometries are known in detail, a household robot would have to handle objects of many shapes and sizes. In a

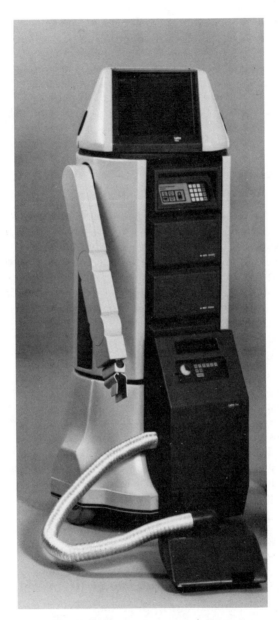

A commercial "home robot" advertised to be a "security sentinel, house cleaner, and personal butler." [Hammacher Schlemmer.]

household environment, these details would change from day to day. Thus, for the household use of robots to become practical, the level of robot command languages and sensing will have to be raised far beyond the requirements of industrial robotics. Such a robot would have to be able to handle natural language inputs, implying a considerable understanding of the household environment, greatly improved visual recognition capabilities, and the ability to move easily and safely amid clutter.

Other Future Applications

Although general-purpose household robots appear infeasible at present, it might be possible to design robots to perform a broad spectrum of other social "housekeeping" tasks. For example, robot sweepers for large public spaces, such as streets, sports stadium, railroad stations, airports, etc., are probably within the reach of current technology. Robot lawnmowers, with programmed awareness of the boundaries of a grassy area, also seem feasible. A related possibility is computer-controlled agricultural equipment for automatic plowing, cultivation, and reaping.

Greater mastery of geography and geometry, based on more sophisticated sensing and locomotion, might permit robot mail, package delivery services, and trash collection. Greatly improved visual and other sensors, together with much more highly developed robot programming techniques, might make robot automobile maintenance, and even some degree of repair, possible. Significantly improved tactile and visual sensing procedures, together with additional theoretical understanding of the behavior of such soft materials as cloth, might allow the development of robot tailors, which could use an appropriate vision algorithm to sense the contours of a client's body and, from this, produce a suit of clothes.

By feeding control signals taken from a human activity into a robot control program, it should be possible to develop various semirobot prostheses, for example, sophisticated grippers for the armless.

TRAINING A GENERATION OF ROBOTICISTS

Although it can be expected to draw upon many other branches of computer science, robotics will have a different flavor from any of them because, in robotics, computer science must go beyond the combinatorial and symbolic manipulations that have been its principal concerns until now. To address the problems of robotics, computer scientists will have to confront the geometric, dynamic, and physical realities of three-dimensional space. This confrontation can be expected to generate a great deal of new science, and robotics should be as central to the next few decades of research in computer science as language, compiler, and system-related investigations have been to the past two decades.

As this happens, a significant reorganization of the present computer science curriculum will be necessary, since robotics research will need to make use of a much wider range of classical mathematics and physics than it has been involved with until now. This includes computational algebra, computational geometry, servomechanism theory, mechanics, theories of elasticity and friction, materials science, and manufacturing technology. Mathematics, physics, and engineering departments will be in closer contact with computer science than ever before. The complexity and high content of classical science may make robotics into a recognized professional subspecialty within computer science.

THE SOCIAL IMPACT OF ROBOTS

To the extent that hazardous, repetitive, and menial tasks are taken over by robots, and to the extent that robots contribute to U.S. productivity and international economic competitiveness, we can take satisfaction in this technology; it simply continues (though significantly generalizing) the trend to increased automation that has characterized the Amercan economy for the past hundred years. On the other hand, some cautionary reflections are suggested. The gradual unfolding of the enormous potential of artificial intelligence can be expected to affect profoundly the fundamental circumstances of human existence. Today's robotic developments exemplify this. Although it would be wrong to forget that much challenging technology must still be put in place before robots can reach their full potential, it is nevertheless true that, as robotic science develops, it will progressively reduce the labor used in industrial production, ultimately to something rather close to zero. If society can respond appropriately to this deep change, we may profit in the manner foretold by Aristotle: "When the Loom spins by itself, and the Lyre plays by itself, man's slavery will be at an end." If, on the other hand, we fail to adjust adequately, for example, if the growth in service industries that has characterized recent U.S. economic history proves insufficient to absorb all of those persons likely to be forced out of industry by the development and installation of robot equipment, grave social tensions may develop. Assessing these difficulties and dealing wisely with them will be important tasks for both social thinkers and national leaders.

Outlook

Although the sensory and manipulative capabilities of industrial robots and their ability to handle unexpected events are still quite limited, robot technology is strengthening and expanding rapidly. Initial commercial success has been achieved through robot mastery of such commercially important operations as spot welding and spray painting. The microelectronic revolution can also be expected to accelerate the development of robotics by supplying many "miracle chips" embodying advanced sensory and control functions.

While all robots will only come to perform the majority of industrial operations gradually, their potential near-term uses are extensive. The area most favorable for robot application at the moment is the manufacture of items in batches substantial enough for hand manufacture to be undesirable but not large enough to warrant heavy capital outlays on fixed automation. Robotic research is proceeding actively in France, West Germany, Sweden, Japan, and the United States, but Japan leads in the use of robots. Operating industrial robots in 1980 numbered 11,250 in Japan and 4,370 in this country.

Much of the robotic equipment used by industry today is too crude to respond flexibly to changing situations. Much more sophisticated control languages are needed to make industrial robots easy to use. Desirable new robot language features include, for example, statements that can cause a manipulator arm to move between specified positions while automatically avoiding certain obstacles or undesirable configurations. In general, we can expect advances in robotic technology to be reflected by the addition of new statements to robot programming languages.

More sophisticated sensory functions, for example, improved image-analysis software, represent another important direction. These sensory capabilities will need to be integrated with manipulator-control software, to produce useful industrial hand-eye systems.

At present, software for analyzing visual data acquired by robots is not especially reliable and is quite expensive computationally. Improved scene-analysis algorithms are needed if we are to have vision systems that can easily be reprogrammed for a wide range of applications. Deeper understanding of the algorithmic nature of such systems will encourage development of high-performance electronic chips that can handle the required operations very rapidly. As such chips become available, the range of objects that can be inspected economically by robot equipment will expand.

The limited present capabilities of industrial robots indicate that we are far from being able to create general-purpose household robots, which would require sensing and command capabilities far exceeding those needed by industry. Still, it may soon be possible to design robots to perform relatively simple housekeeping tasks, such as sweeping large public spaces or mowing lawns. A related possibility is computer-controlled agricultural equipment, such as plows and cultivators.

Robots can be expected to continue to take over hazardous, repetitive, and menial tasks and to contribute significantly to the progress of industrial automation under way in this country for the past century. Moreover, robot technology can be expected to reduce the industrial labor force very drastically. In the absence of adequate mechanisms for adjusting to this change, severe social tensions may result. Our national leaders face the critical task of assessing and forestalling these developing problems.

BIBLIOGRAPHY

Robert Ayres and Steven Miller. *The Impacts of Industrial Robots*. Pittsburgh: Robotics Institute, Carnegie-Mellon University, 1981.

James L. Nevins and Daniel E. Whitney. "Computer-Controlled Assembly," *Scientific American*, Vol. 238, No. 2 (February 1978), pp. 62–74.

Hiroyuki Yoshikawa et al. *Computer-Aided Manufacturing: An International Comparison*. Washington, D.C.: National Academy Press, 1981.

Reviewers*

1 The Genetic Program of Complex Organisms

Roy J. Britten, California Institute of Technology
Paul Marks, Memorial Sloan-Kettering Cancer Center
Philip A. Sharp, Massachusetts Institute of Technology
Charles Yanofsky, Stanford University

2 The Molecular and Genetic Technology of Plants

Lawrence Bogorad, Harvard University
James F. Bonner, California Institute of Technology
Nina Fedoroff, Carnegie Institution of Washington
André T. Jagendorf, Cornell University
Arthur Kelman, University of Wisconsin
Oliver E. Nelson, Jr., University of Wisconsin
John G. Torrey, Harvard University
Virginia Walbot, Stanford University

3 Cell Receptors for Hormones and Neurotransmitters

Douglas M. Fambrough, Carnegie Institution of Washington
Robert Lefkowitz, Duke University Medical Center
Solomon H. Snyder, The Johns Hopkins University School
    of Medicine
J. R. Tata, National Institute for Medical Research, London

---

*Those listed provided substantive comments on early drafts. Responsibility for the final contents rests with the Committee on Science, Engineering, and Public Policy.

4 Psychobiology

Patricia Goldman, Yale University
Donald R. Griffin, Rockefeller University
Eric R. Kandel, Columbia University College of Physicians and Surgeons
Alvin M. Liberman, Haskins Laboratories
Donald B. Lindsley, University of California, Los Angeles
Mark R. Rosenzweig, University of California, Berkeley

5 Surface Science and Its Applications

Robert Gomer, University of Chicago
John H. Sinfelt, Exxon Research and Engineering Company
Gabor A. Somorjai, University of California, Berkeley
William E. Spicer, Stanford University
Maurice Barnett Webb, University of Wisconsin

6 Turbulence in Fluids

Stanley Corrsin, The Johns Hopkins University
Thomas J. Hanratty, University of Illinois at Urbana–Champaign
Mark Kac, University of Southern California
Hans W. Liepmann, California Institute of Technology
Paul Martin, Harvard University
Roger Lyndon Simpson, Southern Methodist University

7 Lasers

Richard G. Brewer, IBM Research Laboratory, San Jose
Henry Kressel, RCA Laboratories
Arthur L. Schawlow, Stanford University

8 The Next Generation of Robots

Mark Fox, Carnegie-Mellon University
Peter Will, Schlumberger Well Services

Index

for the immediacy of the spiritual, transcendent dimension of life./3/ People's religion and apocalyptic are statements against the heavy burden of evolutionary time used by those in power to enslave the people. But whereas the protest has most often been read as one against an institutional presence and for a *spatial* immediacy (the nearness of God, Christ, love, the inner self, nirvana, i.e., immediacy in the spiritual, transcendent dimension of life has been reduced to the immanent), the protest ought also to be read as one against evolutionary presence and for *temporal* immediacy (the nearness of the radically new, i.e., the imminent). What is called for in apocalyptic is indeed a de-historicizing of the future, when that future is merely the *futurum*, the rational extrapolation of existing trends. For that kind of future only de-historicizes the present by robbing it of the possibility for the sudden coming of the new. All that remains in the present is the suffering, impotence, and futurelessness of humankind. A simple reliance on rational extrapolation does not create the future, it destroys it as a dimension of the present. And by destroying it, it not only de-historicizes the future, it de-historicizes the present. It leaves us, as many others have commented, at the brink of (if not in) the post-historical era.

All this is why the "we are not alone" model in science fiction is of such critical importance. For it suggests not merely another model of the science fiction story (thus, criticism of the writer's craft); it suggests a whole new way of reading science fiction. It suggests that science fiction be read not only as a medium for the expression of new ideas and/or scientific data in a palatable form (although it can be that), nor only as a tool for criticizing the present and/or visions of the future (it can be that, too); but it also suggests that science fiction be read as the horizon within which people live and act socially and politically. The former two ways of criticizing science fiction, we repeat, focus on the writer's craft; the latter on the readers' attitudes. Or to put it another way: the former two try to isolate what stands behind the text (as the author stands, literally, behind it); while the latter is more concerned with what stands in front of the text (as the reader does).

This last description also has the advantage of suggesting that the only way to get at the meaning out front is through a critical analysis of the readers. Thus again we are in debt to popular literature, for it is a literature eminently concerned with the reader. It is a user-oriented literature, in the terminology of Herbert Gans. This orientation, moreover, concurs with the analysis of the theory/praxis relationship which has been the focus of much recent systematic (political) theology. Of particular interest is the opposition between those theologians who explicate human experience according to the disclosure models of limit-experience and/or transcendental experience, and those who are concerned with the very structure and meaning of lived human experience insofar as it has a transforming value (Lamb:1978b). The praxis of reading is not valuable merely because it discloses the meaning (theory) intended by the author. Reading itself has a transforming value. There is in this a parallel to the search for the meaning behind the text (i.e., the meaning already present, explicitly or implicitly) as opposed to the meaning out front, which cannot be appropriated except by a participation in the experience of reading.

Finally, we might note that the phrase, "we are not alone," has a double meaning. The "other" is also the other believer. The statement is a testimonial about "we" not "I." There is in this belief ("we are not alone") a proclamation not only that the future as extrapolation is not enough, but also that the present as a private experience is not enough. We see on further reflection that the two are indeed linked: that the most efficient means for political and social oppression is to reduce all experience to private experience, to cut people off not only from the future, but also from each other. This has led us to suggest above that in addition to the desires Ted Peters claims the UFO experiences symbolize (a desire for transcendence, perfection, omniscience, and redemption) we ought also to add the desire for community, for shared faith.

This desire for community and for shared faith leads us in our concluding analysis to the centrality of story and the primacy of narrative theology, both of which are necessary in the discussion of the relation between emancipation and

redemption. Indeed, the correlation of the two foundational
experiences recapitulates the main focus of the present study:
the dialectic of religious and secular soteriologies. Having
established the communal, shared nature of a faith whose desire
lays it open to the coming of the radically new, the question
now is whether this faith can in any way shed light on the
seemingly unbridgeable gulf lying between emancipation and
redemption. And if so, it remains to see how narrative and
story are critical to the enterprise.

Emancipation and Redemption

If there is any formulation of the problematic cutting
across all facets of the concern of the contemporary theological
enterprise to correlate critically human experience and Christian faith, it is Metz's formulation of it in terms of the
relationship between emancipation and redemption. Human experience's current controlling metaphor is emancipation; while
theology's is (and always has been) redemption. Thus we have
here a new perspective on the problem of critical correlation.
But, Metz insists, the critical correlation is not a method
which correlates question and answer, systematic theology and
historical theology, history and dogma; but rather one which
correlates theory and practice, the understanding of faith and
social practice. Emancipation is not the question to which
redemption is the answer; nor, on the other hand, is redemption
the earlier understanding of emancipation. There can be no
baptizing of emancipation; nor can there be a "theological
foothold in the crevices of this dialectic of emancipation"
(1977:323). There can be no argumentative soteriology which
tacks God onto the human process of emancipation, which says,
in effect, in the future we will see the two processes as one.
To do so is to take away from both the understanding of emancipation and the understanding of redemption: for both must be
accepted as mediating a totality, as telling the whole story.
Nor, Metz continues, can there be any compromising theologies
which attempt to assimilate this emancipatory totality, such as
the two-kingdom paradigm or the paradoxical soteriology of the
transcendentalist or existentialist theologies, respectively.

These attempts leave the resulting theological reflection un-
critically aware of the totalizing (and incipient totalitarian)
nature of emancipatory history.

It is precisely this latter concern which leads Metz to
discuss the correlation of emancipation and redemption "within
the framework of what I term...'the history of human suffering'"
(324). For it is precisely insofar as emancipatory history has
failed to come to grips with human suffering (and indeed has
contributed to it as the result of that failure) that a discussion of suffering provides a context in which to speak of
redemption. Suffering and the (hi)stories of suffering raise
first of all the question of guilt. Emancipation, conceived in
terms of a universal historical totality--i.e., humankind as
the subject of history--soon retreats into the "exonerating
mechanism" of abstractions when faced with the question of
guilt for suffering. The subject soon becomes again an object;
this time not of God's actions, but of the processes of technological and economic progress. The result is a "painfully
camouflaged heteronomy" (327). Anthropodicy has replaced
theodicy. A Christian soteriology, on the contrary, does not
flee into abstractions, but demands participation in guilt, in
the concrete praxis of life. Such a participation opens one to
the possibility of redemption not as a process confirming
innocence, but rather one which redeems us from complicity.

Suffering and the (hi)stories of suffering also raise
the question of "deadly finitude"--i.e., what to do not only
with the dead but with death itself. Emancipation justifies
the dead who have died to make us free; but it does not help
the living who still face the same finitude. Thus the cynicism
with regard to the dead whose death has been meaningful, when
compared to the living whose death is still perceived as a
negation and nothing more. Emancipatory progress becomes an
ideology to which humankind is sacrificed generation after
generation. A Christian soteriology, on the contrary, offers
no simple justification of the dead, and so faces head on the
finitude of (our) death and the meaning of (our) freedom. Metz
sees in Jesus' "descent into hell" an example of this--insofar
as it prevents a history of redemption from being reduced to
emancipatory progress looking only to the future. The history